Lecture Notes in Computer Science 2861
Edited by G. Goos, J. Hartmanis, and J. van Leeuwen

Springer

Berlin
Heidelberg
New York
Hong Kong
London
Milan
Paris
Tokyo

Christian Bliek   Christophe Jermann
Arnold Neumaier (Eds.)

Global Optimization and Constraint Satisfaction

First International Workshop on Global Constraint Optimization
and Constraint Satisfaction, COCOS 2002
Valbonne-Sophia Antipolis, France, October 2-4, 2002
Revised Selected Papers

Springer

Series Editors

Gerhard Goos, Karlsruhe University, Germany
Juris Hartmanis, Cornell University, NY, USA
Jan van Leeuwen, Utrecht University, The Netherlands

Volume Editors

Christian Bliek
ILOG, 1681-HB2 Route des Dolines, 06560 Valbonne, France
E-mail: bliek@ilog.fr

Christophe Jermann
Laboratoire IRIN, Université de Nantes,
2, rue de la Houssiniére, BP 44322 Nantes, France
E-mail: christophe.jermann@irin.univ-nantes.fr

Arnold Neumaier
Institute for Mathematics, University Wien, Strudlhofgasse 4, A-1090 Wien, Austria
E-mail: Arnold.Neumaier@univie.ac.at

Cataloging-in-Publication Data applied for

A catalog record for this book is available from the Library of Congress.

Bibliographic information published by Die Deutsche Bibliothek
Die Deutsche Bibliothek lists this publication in the Deutsche Nationalbibliografie;
detailed bibliographic data is available in the Internet at <http://dnb.ddb.de>.

CR Subject Classification (1998): G.1.6, D.3.2-3, I.2.3

ISSN 0302-9743
ISBN 3-540-20463-6 Springer-Verlag Berlin Heidelberg New York

Springer-Verlag is a part of Springer Science+Business Media GmbH

springeronline.com

© Springer-Verlag Berlin Heidelberg 2003
Printed in Germany

Typesetting: Camera-ready by author, data conversion by DA-TeX Gerd Blumenstein
Printed on acid-free paper   SPIN: 10967225   06/3142   5 4 3 2 1 0

Preface

Continuous constraints are the natural way of representing many practical problems and the knowledge they involve. Such constraints may be simple or complex, linear or nonlinear, and may or may not involve transcendental functions. They are widely used to express chemical or mechanical models, process descriptions, building codes or cost restrictions, for example. Many industrial problems involving continuous constraints can be modelled as continuous constraint satisfaction (CSP) or optimization problems (CSOP, often also called mathematical programs).

In practice, such models are often large in size and nonlinear. There may also be constraints involving integer variables, giving rise to mixed integer nonlinear programs (MINLP). The nonlinearities often result in the presence of multiple solutions, or suboptimal local extrema. The challenge is to find all the solutions of a CSP and verify that all have been found, and, in the case of optimization problems, to find the global optimum and verify that it has been found. Complete solution techniques guarantee that all constraints – e.g., safety conditions or tolerance criteria – are satisfied and the global optima identified. Completeness would thus benefit directly the quality and reliability of decisions or analyses based on the provided solutions. This has obvious implications in many industrial and economic areas.

None of the existing approaches for solving nonlinear CSOPs is fully satisfactory in practice. Nonlinear programming techniques are routinely used and can solve large-scale nonlinear problems. However, they are complete only in the convex case and if roundoff errors are controlled. In contrast, constraint programming solvers preserve completeness, but suffer from poor scalability. However, the respective strengths of mathematical programming and constraint programming appear to be highly complementary, and a number of recent developments show that there is a lot to be gained by merging the different inference techniques they provide and by combining their specific advantages.

The goal of the workshop "Global Constrained Optimization and Constraint Satisfaction" (COCOS 2002), which took place during October 2–4, 2002 in Valbonne-Sophia Antipolis (France), was to bring together the communities from global optimization, mathematical programming and constraint programming, giving the participants the opportunity to promote presentation and discussion of ongoing work on solving techniques for continuous CSP and (MI)NLPs. The workshop focused on complete solving techniques for continuous CSOPs which provide all solutions with full rigor. Less rigorous solution techniques were, however, not excluded, since they may be part of complete relevant techniques; for example, local optimization methods are valuable in quickly locating good points and prospective global minimizers.

Three invited lectures and 26 contributed talks were given at the workshop; papers for two of the invited lectures and 15 of the contributed talks are included in the present proceedings.

Invited lectures. The three invited lectures were given by Chris Floudas, Nick Sahinidis, and Baker Kearfott; excluding the author of the constraint satisfaction package Numerica, Pascal Van Hentenryck, who could not attend the workshop, these are the leaders in providing quality software for complete global optimization and constraint propagation. Specifically, Floudas is the author of αBB, and Sahinidis is the author of BARON, both successful global optimization packages based primarily on convex relaxations. Kearfott is the author of GlobSol, an interval-analysis-based global optimization package that provides results with mathematical guarantee, even in the presence of rounding errors.

Unfortunately, Chris Floudas was too busy to provide a written version of his invited lecture *Deterministic Global Optimization: Theoretical, Computational and Implementation Advances*, where he described αBB and its use in industrial applications, mainly in chemical engineering. (Scanned copies of the slides for his talk can be found at [1].) The present volume contains written versions of the remaining two invited lectures.

The paper *Global Optimization and Constraint Satisfaction: The Branch-and-Reduce Approach* by Nick Sahinidis contains a description of the software package BARON for the global solution of mixed integer nonlinear programs, explaining the factorable programming approach, various convex relaxation procedures, and reduction processes based on constraint propagation and Lagrange multiplier techniques. Some typical results for the BARON package (which is complete apart from the possible adverse effects of rounding errors) on real-life applications are presented at the end; the reader is referred to Nick's recent book for more complete descriptions of these applications and results.

The paper *GlobSol: History, Composition, and Advice on Use* by Baker Kearfott describes GlobSol, a rigorous global optimization system incorporating the state of the art in interval analysis, as far as applications to optimization are concerned. Based on automatic differentiation techniques for obtaining interval gradients and Hessians, interval Newton methods for reducing boxes and verifying optimality conditions are discussed. The algorithmic structure is outlined and some advice on how to use the package is given.

Optimization. Eight contributed papers are primarily concerned with optimization. Nowak, Alperin and Vigerske present the object-oriented library LaGO, which solves mixed integer nonlinear programs using convex relaxations. It combines elements from αBB and BARON with convex quadratic underestimation techniques and heuristics for finding feasible points. Henrion and Lasserre show how techniques from semidefinite programming and algebraic geometry can be used to solve many global optimization problems without the need for branching. It is likely that this new approach, which is complementary to the traditional techniques (convexity, intervals, constraint propagation), will play a significant role in the future of global optimization. Jansson shows that the bounds from linear programming (LP) solvers (which are part of the toolkit of all convexity-based global optimization systems) can be made fully rigorous without excessive additional effort, using a simple postprocessing step that is independent of the LP solver used. This raises the possibility of making systems like αBB, BARON and LaGO, which currently suffer from occasional failures due to rounding error issues, fully rigorous. A paper on the COPRIN project (Merlet et al.) shows how to find the exact range for real roots of polynomials with variable coefficients, a global optimization problem with important applications in robotics. Le Thi discuss the reformulation of several classes of optimization problems (including bilevel linear programs) as difference of convex functions (D.C.) programming problems, thus making them amenable to solution with D.C. optimization algorithms. Petrov presents a heuristic for finding good feasible points, relevant for obtaining good upper bounds in global minimization problems, and thus speeding up the branching part of a complete global search. Schmied solves an engine calibration problem by means of a stochastic multistart global optimization algorithm. Mostafa, Vicente and Wright discuss the identification of relevant constraints for degenerate solutions of optimization problems. While the techniques are local they may be of relevance to global problems since these have special difficulties (excessive branching) at degenerate solutions.

Constraint satisfaction. Five contributed papers are primarily concerned with constraint satisfaction problems. Boddy and Johnson show that even large systems of linear and quadratic constraints (with thousands of variables and up to 140,000 constraints) arising in oil refinery problems can be solved successfully using specially adapted constraint propagation and subdivision stategies. Jaulin gives another successful large-scale application, the state estimation of a satellite, using in place of the traditional (local) Kalman filter estimation an interval-based formulation. Walster and Hansen show how to use interval Newton methods to solve overdetermined equations, when the error in the right-hand side is given by worst-case bounds. Cruz and Baharona discuss higher-order consistency methods for constraint-propagation, and argue that maintaining global hull consistency is an efficient alternative. Vu, Sam-Haroud and Silaghi discuss the problem of representing the solution set of constraint satisfaction problems with a continuum of solutions, where a compact description is essential for an intelligible interpretation of the results.

Benchmarking. Finally, two contributed papers discuss questions related to the comparison of different solvers. Shcherbina et al. describe a benchmarking suite containing over 1000 global optimization and constraint satisfaction problems, coded in the AMPL modeling language, and publicly available (with annotations) on the Web. A benchmarking protocol for comparing results on the benchmarking suite is also proposed. Bussieck et al. discuss the assurance of quality in global optimization codes in the GAMS modeling system environment, and associated efforts for standardizing inexpensive, efficient and reproducible tests. They describe the data collection and analysis tools offered by GAMS, which is applicable to the analysis of solver results whether obtained within or outside the GAMS environment.

The present volume of contributions to global optimization and constraint satisfaction shows that the field is growing towards maturity, becoming more reliable and better scalable to larger problems while still posing numerous challenges for the future.

July 2003

Christian Bliek
Christophe Jermann
Arnold Neumaier

References

1. C. Floudas, Deterministic Global Optimization: Theoretical, Computational and Implementation Advances. Slides (132 MB)
   http://liawww.epfl.ch/Cocos02/invited_talks/Slides_Talk_Floudas.pdf

Organization

The COCOS 2002 workshop was organized by the partners of the COCONUT project (IST-2000-26063) with financial support from the European Commission and the Swiss Federal Education and Science Office (OFES).

Program Committee

Frédéric Benhamou	Université de Nantes, France
Christian Bliek	ILOG, France
Boi Faltings	École Polytechnique Fédérale de Lausanne, Switzerland
Arnold Neumaier	University of Vienna, Austria
Peter Spellucci	Darmstadt University, Germany
Pascal Van Hentenryck	Brown University, USA
Luis N. Vicente	University of Coimbra, Portugal

Referees

C. Adjiman	B. Faltings	G. Mayer
C. Audet	F. Goualard	A. Neumaier
F. Benhamou	L. Granvilliers	P. Spellucci
C. Bliek	L. Jaulin	P. Van Hentenryck
M. Ceberio	O. Lhomme	L. Vicente

Table of Contents

Constraint Satisfaction

Benchmarking

Global Optimization and Constraint Satisfaction: The Branch-and-Reduce Approach

Nikolaos V. Sahinidis

University of Illinois, Department of Chemical and Biomolecular Engineering
Urbana, IL 61801, USA
nikos@uiuc.edu
http://archimedes.scs.uiuc.edu

Abstract. In the early 1990s, we proposed the integration of constraint programming and optimization techniques within the branch-and-bound framework for the global optimization of nonconvex nonlinear and mixed-integer nonlinear programs. This approach, referred to as *branch-and-reduce*, was subsequently supplemented with a variety of branching and bounding schemes. In this paper, we review the theory and algorithms behind branch-and-reduce, its implementation in the BARON software, and some recent successful applications.

1 Introduction

The integration of constraint programming and mathematical programming techniques has generated quite some excitement in the operations research and computer science communities in recent years (cf. [16]). Combining these techniques has been found necessary for the solution of many hard combinatorial optimization problems. In the context of nonlinear programs (NLPs) and mixed-integer nonlinear programs (MINLPs), Ryoo and Sahinidis [30, 31] proposed the first variant of the branch-and-reduce algorithm. This algorithm relied on constraints, interval arithmetic, and duality to draw inferences regarding ranges of integer and continuous variables in an effort to expedite the traditional branch-and-bound algorithm for the global optimization of NLPs and MINLPs. Subsequently, this approach was supplemented with branching schemes that lead to finite search trees while branching in continuous spaces [37, 2], and a number of convexification techniques for the construction of relaxations that enjoy tightness along with robustness and computational efficiency [41, 42, 32, 43]. This methodology has been implemented in the computational system BARON [34] and used in a variety of applications, including chemical process design and operation [30, 20], chip layout and design [10], design of just-in-time manufacturing systems [15], optimization under uncertainty [19, 2], pooling and blending problems [1, 44], and molecular design [35, 36].

In Section 2 of this paper, we state the general mixed-integer nonlinear program addressed by this line of research and review algorithms for its solution. Section 3 describes the theoretical and algorithmic components of the branch-and-reduce approach. The implementation is discussed in Section 4, followed by selective computational results in Section 5. Conclusions are drawn in Section 6.

C. Bliek et al. (Eds.): COCOS 2002, LNCS 2861, pp. 1–16, 2003.

2   Mixed-Integer Nonlinear Programming

We address the problem of finding a globally optimal solution of the following mixed-integer nonlinear program:

$$
\begin{array}{ll}
(P) & \min\ f(x) \\
& \text{s.t.}\ \ g(x) \le 0 \\
& \qquad x_i \in \mathbb{R}, \quad i = 1, \ldots, n_\mathrm{d} \\
& \qquad x_i \in \mathbb{Z}, \quad i = n_\mathrm{d} + 1, \ldots, n
\end{array}
$$

where $f : \mathbb{R}^n \mapsto \mathbb{R}$ and $g : \mathbb{R}^n \mapsto \mathbb{R}^m$.

As special cases of P, one recognizes the classical nonlinear and mixed-integer linear programs. Since both of these problem classes are NP-hard [25, 27], it follows that P is also NP-hard. Yet, P is of interest in a large number of applications, including chemical process synthesis, supply chain management and operation, and molecular design [26, 5, 44].

Initial attempts at the solution of P dealt mostly with problems that are convex when integrality restrictions are dropped [14, 11, 6, 7]. A few recent works have indicated that the application of deterministic branch-and-bound algorithms to the global optimization of general classes of MINLPs is promising [30, 39, 13, 38, 44]. Initially developed in the context of combinatorial optimization problems [18, 9], branch-and-bound was later extended to the more general multi-extremal problem P [12, 17]. To solve P, branch-and-bound computes lower and upper bounds on the optimal objective function value over successively refined partitions of the search space. Partition elements are generated and placed on a list of open partition elements. Elements from this list are selected for further processing and further partitioning, and are deleted when their lower bounds are no lower than the best known upper bound for the problem.

Contrary to the pure integer case, finite termination of this algorithm with an exact global optimum of P is, in general, not guaranteed when branching occurs in continuous variable spaces. The algorithm is convergent under fairly general assumptions on the problem functions [17] and, hence, finitely terminating when an ϵ-optimal solution is acceptable with $\epsilon > 0$. Another challenge associated with solving P with branch-and-bound is the construction of lower bounds. While it is straightforward to obtain a relaxation of a mixed-integer linear program by dropping integrality conditions, the development of lower bounds for P requires the (partial) convexification of functions of continuous variables. A further potential difficulty in solving P stems from the presence of nonlinearities. While LP technology has yielded robust and efficient software codes capable of solving large-scale LPs, only much smaller convex NLPs can be solved reliably. Finally, from the practical point of view, acceptance of MINLP algorithms by practitioners requires integration of these algorithms with modeling languages, such as AMPL and GAMS, and necessitates the development of new language concepts that go well beyond the realm of traditional LP and MILP. The next section addresses some of these challenges.

3 Theoretical and Algorithmic Elements of the Branch-and-Reduce Approach

3.1 Factorable Programming Relaxations

In this subsection, we consider the case of factorable functions f and g, *i.e.*, functions that are recursive sums and products of univariate functions. This class of functions suffices to describe most application areas of interest [24]. Examples of factorable functions include $f(x, y) = xy$, $f(x, y) = x/y$, $f(x, y, z, w) = \sqrt{\exp(xy + z \ln w)z^3}$, $f(x, y, z, w) = \left(x^2 y^{0.3} z\right)/w^2 + \exp\left(x^2 w/y\right) - xy$, $f(x) = \sum_{i=1}^{n} \ln_i(x_i)$, and $f(x) = \sum_{i=1}^{T} \prod_{j=1}^{p_i} \left(c_{ij}^0 + c_{ij}x\right)$ with $x \in \mathbb{R}^n$, $c_{ij} \in \mathbb{R}^n$ and $c_{ij}^0 \in \mathbb{R}$ $(i = 1, \ldots, T; j = 1, \ldots, p_i)$.

In his seminar work, McCormick [23] developed bounding techniques for factorable programs. These techniques currently play a central role in many branch-and-bound implementations for problem P. The main observation is that a factorable NLP can be converted to an equivalent separable NLP after the recursive introduction of new variables and constraints [22]. The separable NLP can be relaxed through suitable under- and overestimators of the univariate functions involved. For example, the function $f(x, y, z, w) = \sqrt{\exp(xy + z \ln w)z^3}$, can be decomposed into an almost separable formulation as follows:

$$
\begin{aligned}
x_1 &= xy & x_5 &= \exp(x_4) \\
x_2 &= \ln(w) & x_6 &= z^3 \\
x_3 &= zx_2 & x_7 &= x_5 x_6 \\
x_4 &= x_1 + x_3 & f &= \sqrt{x_7}
\end{aligned}
$$

It is straightforward to outer-approximate the univariate functions $\ln$, $\exp$, and $\sqrt{}$ over bounded intervals. Bilinear terms can be outer-approximated through their convex and concave envelopes over a rectangle [23, 3]:

$$
x_1 \geq \text{convenv}_{[x^L, x^U] \times [y^L, y^U]} = \max\{y^L x + x^L y - y^L x^L, y^U x + x^U y - y^U x^U\}
$$
$$
x_1 \leq \text{concenv}_{[x^L, x^U] \times [y^L, y^U]} = \min\{y^L x + x^U y - y^L x^U, y^U x + x^L y - y^U x^L\}
$$

Several variants of the factorable approach exist. For example, additional variables need not be explicitly introduced. In addition, the decomposition need not proceed until the problem becomes entirely separable. For the example above, non-separabilities in terms of bilinearities were retained. In general, it suffices to proceed only to the extent that the resultant problem can be outer-approximated by a convex feasible set.

3.2 Convexification via Convex Extensions

While factorable programming techniques lead to a completely automatable procedure for the construction of convex lower bounding problems for nonconvex NLPs and MINLPs, these bounding problems often exhibit a large relaxation gap. From the point of view of relaxation quality, it is always advantageous to convexify the original problem functions and constraints to the extent possible. The theory of convex extensions of Tawarmalani and Sahinidis [43] provides a systematic methodology for constructing the closed-form expression of convex envelopes of multidimensional, lower semi-continuous (l.s.c.) functions. This theory provides the capability to construct the convex and concave envelopes of continuous functions. In the sequel, $\bar{\mathbb{R}}$ denotes $\mathbb{R} \cup \{+\infty\}$.

Definition 1 ([43]). *Let C be a convex set and $X \subseteq C$. A convex extension of a function $\phi : X \mapsto \bar{\mathbb{R}}$ over C is any convex function $\eta : C \mapsto \bar{\mathbb{R}}$ such that $\eta(x) = \phi(x)$ for all $x \in X$.*

In other words, an extension is a convex function that agrees with a given (nonconvex) function at each of a predetermined set of points in the domain of definition of both functions. Depending on the original function and this set of points, a convex extension may or may not exist. Furthermore, when a convex extension exists, it need not be unique. In [43], we provide necessary and sufficient conditions for the constructibility of convex extensions.

Key to the development of convex and concave envelopes is the observation that these envelopes are often generated by finitely many sets of points. For instance, in the case of a concave univariate function over an interval, knowledge of the function values at the two endpoints suffices to completely characterize the convex envelope of the function over the interval. This envelope is nothing else but the tightest convex extension of the concave function restricted to the two endpoints. In general, we will refer to this restricted set of points as the *generating set*. One needs to be able to identify the tightest convex extension over this set in order to construct the convex envelope. Working with the convex hull of the epigraph of the convex envelope, allows the latter construction to be easily achieved through disjunctive programming techniques [28], thus leading to the convex envelope in closed-form. This discussion suggests the following two-step procedure for the construction of the convex envelope of a given function:

1. Identify the generating set.
2. Use disjunctive programming to construct the envelope in closed-form.

The main question then becomes how to identify the generating set. We restrict the discussion to functions with compact domains. Let $f(x)$ be the convex envelope of $\phi(x)$ over C, and let F be the epigraph of f. We will use vert(F) to denote the vertex set of F. Then, the convex envelope of ϕ over C is completely specified by the following set:

$$G_C^{\text{epi}}(\phi) = \left\{ x \mid (x, f(x)) \in \text{vert}(F) \right\}.$$

This set is the generating set of the epigraph of function ϕ. The following result characterizes points that do not belong to the generating set:

Theorem 1 ([43]). *Let $\phi(x)$ be a l.s.c. function on a compact convex set C. Consider a point $x_0 \in C$. Then, $x_0 \notin G_C^{\text{epi}}(\phi)$ if and only if there exists a convex subset X of C such that $x_0 \in X$ and $x_0 \notin G_X^{\text{epi}}(\phi)$. In particular, if for an*

ϵ-*neighbourhood* $N_\epsilon \subset C$ *of* x_0 *it can be shown that* $x_0 \notin G_{N_\epsilon}^{\mathrm{epi}}(\phi)$, *then* $x_0 \notin G_C^{\mathrm{epi}}(\phi)$.

We now illustrate the above process in the context of the multilinear function $\phi(x) = \sum_t a_t \prod_{i=1}^{p_t} x_i$, where $-\infty < l_i \le x_i \le u_i < \infty$ for $i = 1, \ldots, n$. When all but one of the n variables are fixed, one is left with a line segment over which ϕ is linear. Hence, only the two endpoints of the line segment may belong to the generating set of the convex as well as concave envelope of ϕ. Applying this argument recursively to all variables and using Theorem 1, it follows [43] that the generating set of the convex as well as concave envelope of ϕ is nothing else but the set of extreme points of the hyperrectangle. Let these points be denoted by p_k, $k = 1, \ldots, 2^n$, and let ϕ_k be the corresponding values of ϕ at these points. The polyhedral description of the convex outer-approximator of ϕ follows trivially from polyhedral representation theorems (cf. Theorem 4.8, p. 96 in [27]):

$$x = \sum_{k=1}^{2^n} \lambda_k p_k, \quad \phi = \sum_{k=1}^{2^n} \lambda_k \phi_k, \quad \sum_{k=1}^{2^n} \lambda_k = 1$$
$$\lambda_k \ge 0, \quad k = 1, \ldots, 2^n$$

The above approach can be readily applied to obtain the convex and concave envelopes of functions of the following forms:

- $\phi(x, y) = M(x_1, x_2, \ldots, x_n)/(y_1^{a_1}, y_2^{a_2}, \ldots, y_m^{a_m})$ over a hyperrectangle, where M is a multilinear expression, $y_1, \ldots, y_m \ne 0$, and $a_1, \ldots, a_m \ge 0$. For example, consider $(x_1 x_2 + x_3 x_4)/(y_1 y_2 y_3)$.
- $\phi(x, y) = f(x) \sum_{i=1}^n \sum_{j=-p}^k a_{ij} y_i^j$ over a hyperrectangle, where f is concave, $a_{ij} \ge 0$ for $i = 1, \ldots, n; j = -p, \ldots, k$, and $y_i > 0$. For example, consider $x/y + 3x + 4xy + 2xy^2$.

Additional examples and generalizations of this methodology can be found in [42, 43]. In [44], the theory of convex extensions was used to show that a particular relaxation of the pooling problem dominates earlier ones in the literature. An additional application of this theory is described next.

3.3 Convexification via Product Disaggregation

Throughout this subsection, we consider the following function:

$$\phi(x; y_1, \ldots, y_n) = a_0 + \sum_{k=1}^n a_k y_k + x b_0 + x \sum_{k=1}^n b_k y_k$$

where a_k and b_k $(k = 0, \ldots, n)$ are given constants, $x \in [x^L, x^U]$, and $y_k \in [y_k^L, y_k^U]$. The convex extensions theory can be used to prove the following result:

Theorem 2 ([40]). *Let* $H = [x^L, x^U] \times \prod_{k=1}^n [y_k^L, y_k^U]$. *Then:*

$$\mathrm{convenv}_H \, \phi = a_0 + \sum_{k=1}^n a_k y_k + x b_0 + \sum_{k=1}^n \mathrm{convenv}_{[y_k^L, y_k^U] \times [x^L \times x^U]}(b_k y_k x).$$

The standard way to bound ϕ is to substitute w for $\sum_{k=1}^{n} b_k y_k$ and then relax xw using the bilinear envelopes of Subsection 3.1. The required bounds on w follow by maximizing/minimizing $\sum_{k=1}^{n} b_k y_k$ over $\prod_{k=1}^{n} [y_k^L, y_k^U]$. While commonplace, such a construction does not yield the convex envelope of $\phi(x, y)$. Consider, for example, $x(2y_1 + 3y_2)$ over $[0, 1]^3$. At $x = 0.5$, $y_1 = 1$, and $y_2 = 0$, the convex envelope equals $x(2y_1 + 3y_2)$ with a value of 1 whereas the standard lower bounding procedure gives a value of 0. Just like the standard lower bounding procedure, the convex envelope of Theorem 2 also makes use of the bilinear convex envelope formula of Subsection 3.1. However, the bilinear envelopes are invoked only after the product is distributed over the summation.

Distribution of the product over the summation results in disaggregating xw into $\sum_{k=1}^{n} b_k z_k$, where $z_k = xy_k$. This procedure was termed *product disaggregation* in [40] as it is reminiscent of *variable disaggregation*, a procedure that provides tight linear relaxations of mixed-integer linear programs [27]. In [40], we provide a number of applications of product disaggregation, including fractional programs and certain optimization problems that arise from the discretization of dynamic systems.

Theorem 3 ([40]). *Let $f(x; y_1, \ldots, y_n)$ be the convex envelope of $\phi(x; y_1, \ldots, y_n)$ over $[x^L, x^U] \times \prod_{k=1}^{n} [y_k^L, y_k^U]$ and let $f_r(x; y_1, \ldots, y_n)$ be the convex envelope of $\phi(x; y_1, \ldots, y_n)$ over $[x^L, x_r^U] \times \prod_{k=1}^{n} [y_k^L, y_k^U]$, where $x_r^U < x^U$. Let $K^+ = \{k \mid b_k > 0\}$ and $K^- = \{k \mid b_k < 0\}$. Then, $f_r(x^0; y_1^0, \ldots, y_n^0) > f(x^0; y_1^0, \ldots, y_n^0)$ if and only if $(x^0; y_1^0, \ldots, y_n^0) \in S$ where S is given by:*

$$S = \bigcup_{k \in K^+} \left\{ (x, y) \mid (y_k^U - y_k^L)x + (x_r^U - x^L)y_k - x_r^U y_k^U + x^L y_k^L > 0, y_k < y_k^U \right\} \cup \bigcup_{k \in K^-} \left\{ (x, y) \mid (y_k^U - y_k^L)x + (x^L - x_r^U)y_k + x_r^U y_k^L - x^L y_k^U > 0, y_k > y_k^L \right\}.$$

A similar result is presented in [40] when the lower bound on x is improved. These results highlight the importance of reducing bounds on x as much as possible when one is interested in deriving convex outer-approximators.

3.4   Polyhedral Outer-Approximation

The current state-of-the-art in linear programming permits the reliable solution of very large-scale LPs in reasonable computational times. On the contrary, nonlinear programs are harder to solve. As a result, it is often advantageous to use polyhedral instead of other convex relaxations in branch-and-bound, even when the latter relaxations are tighter. For this reason, Tawarmalani and Sahinidis [41] developed a polyhedral outer-approximation scheme that generates an entirely linear programming relaxation. The starting point of this approach is a convex nonlinear relaxation obtained by factorable programming and/or convex extensions techniques. Subsequently, a sandwich algorithm is used to provide a polyhedral outer-approximation of the nonlinear functions.

The sandwich algorithm is a template of outer-approximation schemes [8, 29]. At a given iteration, this algorithm begins with a number of points at which tangential outer-approximations of the convex function have been constructed. Then, at every iterative step, the algorithm identifies the interval with the maximum outer-approximation error and subdivides it at a suitably chosen point.

Let $\mathcal{G}(f)$ denote the set of points (x, y) such that $y = f(x)$. Let $\phi^o(x)$ be the outer-approximation of $\phi(x)$, and consider the projective error measure:

$$\epsilon_p = \sup_{p^{\phi^o} \in \mathcal{G}(\phi^o)} \inf_{p^\phi \in \mathcal{G}(\phi)} \left\{ \| p^\phi - p^{\phi^o} \| \right\}.$$

Given a number of outer-approximators, the strategy developed in [41] constructs the next outer-approximator as the line supporting ϕ at a point where the maximum projective error occurs. This scheme converges quadratically:

Theorem 4 ([41]). *Consider the univariate function* $\phi : [x^L, x^U] \mapsto \mathbb{R}$. *Let* $R = \left(x^L, \phi(x^L) \right)$ *and* $S = \left(x^U, \phi(x^U) \right)$. *Assume that the tangents at R and S intersect at O. Let θ be $\pi - \angle ROS$, and $L = |RO| + |OS|$. Let $k = L\theta/\epsilon_p$, where ϵ_p is the maximum allowable projective error. Then, the algorithm needs at most* $\lceil \sqrt{k} - 2 \rceil$ *supporting lines to achieve the required accuracy.*

3.5 Branch-and-*Reduce*

The previous subsections have illustrated that the quality of the relaxations thus obtained is a strong function of the bounds of variables that participate in nonlinear relationships. Thus, tighter variable bounds imply tighter relaxations and can be expected to lead to faster convergence of branch-and-bound. Our approach to global optimization places a strong emphasis on the derivation of tight bounds for all problem variables. In each node of the search tree, constraint programming techniques are utilized in a *preprocessing* step to reduce ranges of problem variables before a relaxation is constructed. Once the relaxation is solved, a *postprocessing* step utilizes the solution of the relaxed problem in an attempt to further reduce ranges of variables before branching occurs. Precisely because so much emphasis is placed on the reduction of ranges of problem variables, we refer to the overall algorithm as a branch-and-*reduce* approach. The main reduction strategies used in our framework are outlined next.

3.6 Drawing Inferences from Constraints: Feasibility-Based Range Reduction

Feasibility-based tightening, or feasibility-based range reduction, is a process that relies on the problem constraints to cut-off infeasible portions of the solution space. Assume, for example, that the following constraints are part of the problem to be solved at a given node:

$$\sum_{j=1}^{n} a_{ij} x_j \leq b_i, \quad i = 1, ..., m.$$

To tighten variable bounds based on these linear constraints, one could simply solve the $2n$ LPs:

$$\left\{ \min \pm x_k \text{ s.t. } \sum_{j=1}^{n} a_{ij} x_j \le b_i, i = 1, \ldots, m \right\}, k = 1, \ldots, n, \qquad (1)$$

which would provide tightening that is optimal, albeit computationally expensive. An alternative approach is based on the observation that one of the constraints

$$\begin{cases} x_h \le \frac{1}{a_{ih}} \left(b_i - \sum_{j \ne h} \min \left\{ a_{ij} x_j^U, a_{ij} x_j^L \right\} \right), & a_{ih} > 0 \\ x_h \ge \frac{1}{a_{ih}} \left(b_i - \sum_{j \ne h} \min \left\{ a_{ij} x_j^U, a_{ij} x_j^L \right\} \right), & a_{ih} < 0 \end{cases} \qquad (2)$$

is also valid for each pair (i, h) that satisfies $a_{ih} \ne 0$. The constraints in (2) function as "poor man's linear programs," particularly when they are applied iteratively, looping over the set of variables several times. Constraints (2) have been used extensively in the constraint programming literature [16] as well as in the mixed-integer linear programming literature [4]. There are well-known pathological situations in which these constraints do not provide optimal or even any tightening. Shectman and Sahinidis [37] experimented with these constraints in the context of concave minimization over polytopes demonstrating that these constraints often provide optimal tightening, particularly when the full LPs (1) are solved once at the root node of the search tree. Our current approach is to solve, at every node of the search tree, the full LPs (1) for a few judiciously selected variables and aggressively apply the approximate strategy (2) to all variables.

The above approach can be extended to the case of nonlinear constraints, giving rise to "poor man's nonlinear programs," an approach particularly easy to implement in the context of factorable and separable nonlinear programs.

3.7   Drawing Inferences from Optimal Solutions: Optimality-Based Range Reduction

Optimality-based range reduction recognizes that dual solutions of the relaxation solved at any node of the search tree provide information about the shape of the value function of the relaxed problem. This information can be used to construct an underestimator of the value function of the relaxed problem that, in turn, underestimates the value function of the nonconvex problem. Requiring the so-constructed underestimator to take values below that of the current best known solution, leads to inferences regarding inferior parts of the search space.

In their simplest form, optimality-based inferences can be drawn about variable ranges as shown by Ryoo and Sahinidis [30]. Assume that the simple range constraint $x \le x^U$ is active at a relaxed problem solution with a corresponding

Lagrange multiplier of $\lambda > 0$. Let L and U denote the objective function values of the relaxed problem's solution and the incumbent, respectively. Clearly, then, $L - \lambda(x - x^U)$ provides a first-order underestimator of the relaxation value function. Requiring this underestimator to take better values than U, gives the simplest optimality-based range reduction cut:

$$x \geq x^U - \frac{U - L}{\lambda}.$$

Similar inferences can be drawn about variables that, at the relaxed problem solution, go to their lower bounds. If a variable is at neither of its bounds in the relaxed problem solution, we can *probe* its bounds by temporarily fixing this variable at (or close to) its lower (upper) bound, constructing the linear underestimator of the value function, and contracting the range of the variable using the dual solution thus obtained. This process requires the solution of additional relaxations and, in certain cases, is less advantageous than solving the full range contraction LPs or NLPs discussed in the previous section.

In [30, 31], the above process of range contraction is extended to arbitrary constraints of the type $g(x) \leq 0$, or even to sets of constraints that may or may not be active at the relaxed problem solution. One is then able to infer valid inequalities, some of which may be nonconvex. Although they may exclude solutions that are feasible to P, these inequalities do not exclude any solutions of P with objective function values better than U.

3.8 A Unified Framework for Constraint Inferencing

The range reduction schemes of the two previous subsections involve the solution of some optimization problem. In the case of feasibility-based reduction, an optimization problem such as (1) is solved either approximately or exactly. In the case of optimality-based reduction, one solves a relaxation. Observe that these optimization problems are very closely related. In the simple case of a linearly constrained global optimization problem, the feasible space of these optimization problems is identical. Thus, any feasible dual solution of the relaxation is also dual feasible to the range reduction problem (1) and *vice versa*. Hence, solutions obtained for one problem can be used for range reduction in the other problem. In Chapter 6 of [44], Tawarmalani and Sahinidis build on this observation to provide a unified range reduction theory that subsumes the feasibility-based and optimality-based range reduction schemes of the two previous subsections as well as a variety of such schemes from the literature [21, 4, 30, 31, 46].

3.9 Node Selection and Branching

In contrast to branching on 0−1 variables, branching on continuous variables may not lead to finite partitioning. Consequently, the lower and upper bounding sequences generated by the algorithm are, in general, only convergent in the latter case. To ensure convergence, we use a *bound-improving* node selection rule by selecting a partition element with the current lowest bound every finitely many iterations. In addition, by periodically bisecting the longest edge amongst all nonlinear variables, we ensure an *exhaustive* partitioning scheme, *i.e.*, one that guarantees that partition elements converge to points or other sets over which P is easily solvable. This strategy guarantees convergence of the overall algorithm [17]. We have shown that, when applied to concave minimization over polytopes, this algorithm is finite as long as one branches on the incumbent solution when possible. For two-stage stochastic programs with integer variables in the second stage, we have shown that branching in the space of the "tender variables" renders this algorithm finite for this problem class as well [2].

We rely exclusively on rectangular partitioning. A variable is chosen and its range partitioned at the relaxation solution—unless bisection is required for convergence or branching on the incumbent is possible. Note, however, that when factorable relaxations are used, the problem is reformulated in a higher dimensional space. Branching on the reformulation variables may then result to partitions that are not rectangular in the original problem space. Finally, we note that the process for selecting the branching variable accounts for all deviations of outer-approximators from original nonlinear problem functions for which a variable is responsible for. Further, we account for the current relaxation problem solution and potential for its improvement as detailed in Chapter 6 of [44] in order to compute branching priorities. The variables with the largest branching priorities are considered candidates for probing.

3.10   Finding the K Best or All Feasible Solutions

Consider an optimization problem with k integer variables, x_i, $i = 1, \ldots, k$. For simplicity, we assume $0 \le x_i \le x_i^U$, $i = 1, \ldots, k$. It is common practice to identify multiple solutions of such a problem in one of the two following ways:

– Given a solution x^*, introduce the following *nonlinear* cut:

$$\sum_i (x_i^* - x_i)^2 \ge 1.$$

– Reformulate the problem by introducing binary variables:

$$x_i = \sum_{j=1}^{\lfloor log_2(x_i^U) \rfloor} 2^{j-1} y_{ij}, \quad i = 1, \ldots, k$$

The solution y^* can be excluded by the well-known *linear* integer cut:

$$\sum_{(i,j) \in \mathcal{B}^*} y_{ij} - \sum_{(i,j) \in \mathcal{N}^*} y_{ij} \le |\mathcal{B}^*| - 1$$

where $\mathcal{B}^* = \{(i,j)|y_{ij}^* = 1\}$ and $\mathcal{N}^* = \{(i,j)|y_{ij}^* = 0\}$.

Several solutions of the problem can then be obtained by solving a series of models in which integer cuts are successively introduced to exclude the previous models' optimal solutions from further consideration. However, such an approach requires the search of a number of branch-and-bound trees.

Instead of using the integer cuts above, we modify the standard node fathoming step of the algorithm. Instead of deleting all inferior nodes when a feasible solution is found, we delete only the current node when it becomes infeasible or a point. Nodes where feasible solutions are identified are branched further until they become points in the search space or infeasible. All feasible solutions can be identified through a single application of branch-and-reduce.

Once the fathoming step of the algorithm has been modified as above, the optimality-based range reduction and probing techniques of Subsection 3.7 must also be modified. These techniques require an appropriate upper bound for optimality-based fathoming. Instead of using the upper bounds provided by feasible solutions identified during the search, we make use of a bound obtained by solving a linear relaxation of the problem.

Note that it is straightforward to modify this algorithm to provide only the K best solutions and that this scheme will work well in continuous spaces provided that the sought-after solutions are isolated (separated by a certain distance).

4   The BARON Computational System

The Branch-And-Reduce Optimization Navigator (BARON) implements the algorithms described above by combining branch-and-bound with constraint propagation and duality techniques for reducing ranges of variables in the course of the algorithm. From the very beginning, BARON was developed as a *user-configurable system* that could be easily modified by users to allow experimentation with different lower bounding, branching, and other algorithmic options.

The first version of BARON was merely 1800 lines of code written in the GAMS modeling language in 1991-93. The software evolved into 10,000 lines of FORTRAN 77 in 1994-95. Currently, BARON is a mix of about 42,000 lines in FORTRAN 90 and 24,000 lines in C. It still serves as a system that facilitates experimentation with novel branch-and-bound algorithms for global optimization. In addition, it provides a modeling language and a completely automated way for solving NLPs and MINLPs to global optimality. The latter option is also offered under the GAMS modeling system as illustrated in Chapter 11 of [44]. Other components of the system include an automatic function evaluator and differentiator, sparse matrix utilities, data manager, and links to solvers for the solution of LP, NLP, and SDP subproblems. While no complete rounding error control is currently attempted, an IEEE exception handler has been fully developed for objective and constraint function calculations for local search and lower bounding purposes.

5   Computations

Extensive computations with the proposed algorithm on over 500 problems are reported in [44]. Table 1 presents computational results with selected problems on a 332 MHz RS/6000 Model 43P with 128MB RAM and a LINPACK score of 59.9. For all problems solved, we report the total CPU seconds taken to solve the problem (T_{tot}), the total number of nodes in the branch-and-reduce tree (N_{tot}), and the maximum number of nodes that had to be stored in memory during the search (N_{mem}). Computations were carried out with an absolute termination tolerance (difference between upper and lower bounds) of 10^{-6}.

The problems of Table 1 include pooling problems (adhya1 to adhya4), a non-linear fixed-charge problem (e27), a reliability problem (e29), a mechanical fixture design problem (e31), a heat exchanger network synthesis problem (e35), a pressure design problem (e38), a truss design problem (e40), a problem from the design of just-in-time manufacturing systems (jit), a particulary challenging molecular design problem (primary), and two difficult problems from discretization of dynamical systems (tiny, catmix).

Finally, Figure 1 shows results for the robot problem [45], a set of 8 quadratic equations in 8 unknowns. We utilize BARON's numsol option to identify solutions within an isolation tolerance of 10^{-4}. For numsol = 1, the algorithm requires 10 nodes to obtain the solution. For numsol = 16, 334 nodes are required, i.e., approximately 21 nodes per solution found. For numsol $\geq$ 17, only 16 solutions are obtained thus proving that this problem has exactly 16 solutions.

Table 1. Selected computational results for problems from [44]

Problem	Obj.	m	n	n_d	T_{tot}	N_{tot}	N_{mem}
e27	2.00	2	1	1	0.02	3	2
e40	30.41	7	4	3	0.15	24	6
e29	-0.94	6	2	2	0.18	47	11
e38	7197.73	3	4	2	0.38	5	2
jit	173,983	33	26	4	0.67	63	10
adhya1	-549.80	52	13	0	1.20	5	1
adhya4	-877.65	67	18	0	1.35	1	1
adhya2	-549.80	69	13	0	1.75	11	1
adhya3	-561.05	78	20	0	1.95	5	1
e36	-246.00	2	2	1	2.59	768	72
e31	-2.00	135	112	24	3.75	351	56
e32	-1.43	18	35	19	13.7	906	146
e35	64868.10	39	32	7	16.4	465	57
primary	-1.2880	164	82	58	375	15930	1054
tiny	1.00594	96	71	16	1110	3728	244
catalyst	-0.01637	32	33	0	3540	3477	480

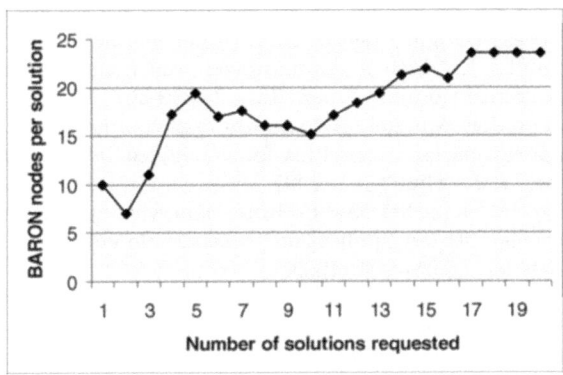

Fig. 1. BARON nodes per solution for the robot problem for different `numsol` values

6   Conclusions

The problems in Table 1 have been sorted in order of increasing time. Looking at this table, one observes that size alone is not a good indicator of problem difficulty. Nor is the number of integer variables. From the results of this table, it is clear that problems with up to a few hundred variables and constraints are solvable with the general-purpose BARON system. In [20, 37, 33, 44], we report computational results on problems with up to a few thousand variables with specialized implementations of branch-and-reduce. The problems of Table 1 are coming from a very wide variety of applications, demonstrating the broad applicability of the algorithms described in this paper.

Acknowledgements

Partial financial support from the National Science Foundation under awards ECS 00-98770, DMI 01-15166, and CTS 01-24751 is gratefully acknowledged.

References

[1] N. Adhya, M. Tawarmalani, and N. V. Sahinidis. A Lagrangian approach to the pooling problem. Industrial & Engineering Chemistry, 38:1956-1972, 1999.

[2] S. Ahmed, M. Tawarmalani, and N. V. Sahinidis. A finite branch and bound algorithm for two-stage stochastic integer programs. Mathematical Programming. Submitted, 2000.

[3] F. A. Al-Khayyal and J. E. Falk. Jointly constrained biconvex programming. Mathematics of Operations Research, 8:273-286, 1983.

[4] D. E. Andersen and K. D. Andersen. Presolving in linear programming. Mathematical Programming, 71:221-245, 1995.

[5] L. T. Biegler, I. E. Grossmann, and A. W. Westerberg. Systematic Methods of Chemical Process Design. Prentice Hall, Upper Saddle River, New Jersey, 1997.

[6] B. Borchers and J. E. Mitchell. An improved branch and bound for mixed integer nonlinear programs. Comput. Oper. Res., 21:359-367, 1994.

[7] B. Borchers and J. E. Mitchell. A computational comparison of branch and bound and outer approximation algorithms for 0-1 mixed integer nonlinear programs. Comput. Oper. Res., 24:699-701, 1997.

[8] R. E. Burkard, H. Hamacher, and G. Rote. Sandwich approximation of univariate convex functions with an application to separable convex programming. Naval Research Logistics, 38:911-924, 1992.

[9] R. J. Dakin. A tree search algorithm for mixed integer programming problems. Computer Journal, 8:250-255, 1965.

[10] M. C. Dorneich and N. V. Sahinidis. Global optimization algorithms for chip layout and compaction. Engineering Optimization, 25:131-154, 1995.

[11] M. A. Duran and I. E. Grossmann. An outer-approximation algorithm for a class of mixed-integer nonlinear programs. Mathematical Programming, 36:307-339, 1986.

[12] J. E. Falk and R. M. Soland. An algorithm for separable nonconvex programming problems. Management Science, 15:550-569, 1969.

[13] C. A. Floudas. Deterministic Global Optimization: Theory, Algorithms and Applications. Kluwer Academic Publishers, Dordrecht, 1999.

[14] O. K. Gupta and A. Ravindran. Branch and bound experiments in convex nonlinear integer programming. Management Science, 31:1533-1546, 1985.

[15] R. A. Gutierrez and N. V. Sahinidis. A branch-and-bound approach for machine selection in just-in-time manufacturing systems. International Journal of Production Research, 34:797-818, 1996.

[16] J. Hooker. Logic-Based Methods for Optimization: Combining Optimization and Constraint Satisfaction. John Wiley & Sons, New York, NY, 2000.

[17] R. Horst and H. Tuy. Global Optimization: Deterministic Approaches. Springer Verlag, Berlin, Third edition, 1996.

[18] A. H. Land and A. G. Doig. An automatic method for solving discrete programming problems. Econometrica, 28:497-520, 1960.

[19] M. L. Liu and N. V. Sahinidis. Process planning in a fuzzy environment. European J. Operational Research, 100:142-169, 1997.

[20] M. L. Liu, N. V. Sahinidis, and J. P. Shectman. Planning of chemical process networks via global concave minimization. In I. E. Grossmann (ed.), Global Optimization in Engineering Design, Kluwer Academic Publishers, Boston, MA, pages 195-230, 1996.

[21] O. L. Mangasarian and L. McLinden. Simple bounds for solutions of monotone complementarity problems and convex programs. Mathematical Programming, 32:32-40, 1985.

[22] G. P. McCormick. Converting general nonlinear programming problems to separable nonlinear programming problems. Technical Report T-267, The George Washington University, Washington D. C., 1972.

[23] G. P. McCormick. Computability of global solutions to factorable nonconvex programs: Part I-Convex underestimating problems. Mathematical Programming, 10:147-175, 1976.

[24] G. P. McCormick. Nonlinear Programming: Theory, Algorithms and Applications. John Wiley & Sons, 1983.

[25] K. G. Murty and S. N. Kabadi. Some NP-complete problems in quadratic and nonlinear programming. Mathematical Programming, 39:117-129, 1987.

[26] S. V. Nabar and L. Schrage. Formulating and solving business problems as nonlinear integer programs. Technical report, Graduate School of Business, University of Chicago, 1992.

[27] G. L. Nemhauser and L. A. Wolsey. Integer and Combinatorial Optimization. Wiley Interscience, Series in Discrete Mathematics and Optimization, 1988.

[28] R. T. Rockafellar. Convex Analysis. Princeton Mathematical Series. Princeton University Press, 1970.

[29] G. Rote. The convergence rate of the sandwich algorithm for approximating convex functions. Computing, 48:337-361, 1992.

[30] H. S. Ryoo and N. V. Sahinidis. Global optimization of nonconvex NLPs and MINLPs with applications in process design. Computers & Chemical Engineering, 19:551-566, 1995.

[31] H. S. Ryoo and N. V. Sahinidis. A branch-and-reduce approach to global optimization. Journal of Global Optimization, 8:107-139, 1996.

[32] H. S. Ryoo and N. V. Sahinidis. Analysis of bounds for multilinear functions. Journal Global Optimization, 19:403-424, 2001.

[33] H. S. Ryoo and N. V. Sahinidis. Global optimization of multiplicative programs. Journal of Global Optimization. Accepted, 2002.

[34] N. V. Sahinidis. BARON: A general purpose global optimization software package. Journal of Global Optimization, 8:201-205, 1996.

[35] N. V. Sahinidis and M. Tawarmalani. Applications of global optimization to process and molecular design. Computers & Chemical Engineering, 24:2157-2169, 2000.

[36] N. V. Sahinidis, M. Tawarmalani, and M. Yu. Design of alternative refrigerants via global optimization. AIChE J. Accepted, 2003.

[37] J. P. Shectman and N. V. Sahinidis. A finite algorithm for global minimization of separable concave programs. Journal of Global Optimization, 12:1-36, 1998.

[38] H. D. Sherali and H. Wang. Global optimization of nonconvex factorable programming problems. Mathematical Programming, 89:459-478, 2001.

[39] E. M. B. Smith and C. C. Pantelides. Global optimisation of general process models. In I. E. Grossmann (ed.), Global Optimization in Engineering Design, Kluwer Academic Publishers, Boston, MA, pages 355-386, 1996.

[40] M. Tawarmalani, S. Ahmed, and N. V. Sahinidis. Product disaggregation and relaxations of mixed-integer rational programs. Optimization and Engineering, 3:281-303, 2002.

[41] M. Tawarmalani and N. V. Sahinidis. Global optimization of mixed-integer nonlinear programs: A theoretical and computational study. Mathematical Programming. Submitted, 1999.

[42] M. Tawarmalani and N. V. Sahinidis. Semidefinite relaxations of fractional programs via novel techniques for constructing convex envelopes of nonlinear functions. Journal of Global Optimization, 20:137-158, 2001.

[43] M. Tawarmalani and N. V. Sahinidis. Convex extensions and convex envelopes of l.s.c. functions. Mathematical Programming, 93:247-263, 2002.

[44] M. Tawarmalani and N. V. Sahinidis. Convexification and Global Optimization in Continuous and Mixed-Integer Nonlinear Programming: Theory, Algorithms, Software, and Applications, volume 65 of Nonconvex Optimization and Its Applications. Kluwer Academic Publishers, Dordrecht, 2002.

[45] L.-W. Tsai and A. P. Morgan. Solving the kinematics of the most general sixand five-degree-of-freedom manipulators by continuation methods. ASME J. of Mechanisms, Transmissions and Automation in Design, 107:48-57, 1985.

[46] J. M. Zamora and I. E. Grossmann. A branch and contract algorithm for problems with concave univariate, bilinear and linear fractional terms. Journal of Global Optimization, 14:217-249, 1999.

GlobSol:
History, Composition, and Advice on Use

R. Baker Kearfott

University of Louisiana at Lafayette, Department of Mathematics
Box 4-1010, Lafayette, LA 70504-1010, USA
rbk@louisiana.edu
Telephone: (337) 482-5270

Abstract. The GlobSol software package combines various ideas from interval analysis, automatic differentiation, and constraint propagation to provide verified solutions to unconstrained and constrained global optimization problems. After briefly reviewing some of these techniques and GlobSol's development history, we provide the first overall description of GlobSol's algorithm. Giving advice on use, we point out strengths and weaknesses in GlobSol's approaches. Through examples, we show how to configure and use GlobSol.

Keywords: Verified global optimization, interval analysis, GlobSol, constraint propagation, automatic differentiation

1 Introduction

Specific forms of the following related problems occur throughout scientific computing:

$$
\boxed{\text{Given a system of equations } F(x) = 0, \text{ find a point } \check{x} \text{ such that } F(\check{x}) = 0,}
\tag{1}
$$

and

$$
\boxed{\begin{aligned}
&\text{minimize } \varphi(x) \\
&\text{subject to } c_i(x) = 0, \ i = 1, \ldots, m_1, \\
&\qquad\qquad\ g_i(x) \leq 0, \ i = 1, \ldots, m_2, \\
&\text{where } \varphi : \mathbb{R}^n \to \mathbb{R} \text{ and } c_i, g_i : \mathbb{R}^n \to \mathbb{R}.
\end{aligned}}
\tag{2}
$$

Numerous algorithms, such as those in [4, 6], use heuristics to find points $\check{x}$ to approximately solve (1) or (2). Although successful and in wide use for a number of practical problems, such algorithms come with no guarantees or error bounds on the resulting approximate solutions $\check{x}$. Indeed, there are instances where approximate solutions $\check{x}$ have been published as true, but have been far from true solutions, and where subsequent rigorous investigation has revealed true solutions to the model. Furthermore, these true solutions happen to be more meaningful physically; see [5].

Thus, although more difficult with regard to both computational complexity and implementation, it is sometimes useful to validate approximate solutions

C. Bliek et al. (Eds.): COCOS 2002, LNCS 2861, pp. 17–31, 2003.

$\check{x}$. For problem(2) (and with related algorithms for problem (1)), this is done with *deterministic global optimization*. In the context of deterministic global optimization, problem (2) becomes[1]

> Given a box $x = ([\underline{x}_1, \overline{x}_1], \ldots [\underline{x}_n, \overline{x}_n])$, find narrow boxes
> $x^* = ([\underline{x}_1^*, \overline{x}_1^*], \ldots [\underline{x}_n^*, \overline{x}_n^*])$ such that any solutions of
>   minimize $\varphi(x)$
>   subject to $c_i(x) = 0$, $i = 1, \ldots, m_1$,
> $g_i(x) \leq 0$, $i = 1, \ldots, m_2$,
> where $\varphi : \mathbb{R}^n \to \mathbb{R}$ and $c_i, g_i : \mathbb{R}^n \to \mathbb{R}$
> are guaranteed to be within one of the x^* that has been found.　　　　　(3)

Deterministic global optimization algorithms, instances of *branch and bound* procedures, contain a method of bounding the ranges of φ, c, and g over x and over sub-boxes of x, combined with some method of subdividing a box x into smaller sub-boxes. In interval-based algorithms, interval arithmetic is used to bound the ranges. Since, with directed rounding, interval arithmetic is rigorous, interval-based branch and bound algorithms give bounds on the solution with mathematical certainty.

2　Elements of GlobSol

Specific interval branch and bound algorithms contain additional techniques to more efficiently eliminate subregions of the original region x that do not contain solutions, and to provide time-saving user interfaces. Such techniques include

1. automatic differentiation,
2. constraint propagation,
3. interval Newton methods,
4. additional, specialized techniques.

The GlobSol software is a Fortran 90-based package for global optimization that uses an interval branch and bound procedure with these additional techniques. Before describe GlobSol's overall algorithm, we provide pointers to the literature for these underlying techniques, along with synopses of how the techniques are used in GlobSol.

2.1　Automatic Differentiation

GlobSol uses automatic differentiation so the user need only input the objective φ and constraints c_i and g_i, without need to worry about programming gradients and Hessian matrices or about programming separate floating point and interval versions, etc. Although various methods are possible, GlobSol uses *operator overloading* in a preprocessing step to first compute an internal, symbolic

---

[1] Throughout, as in formula (3), we denote interval vectors with boldface.

representation of the objective and gradients, which we call a "code list." The actual GlobSol branch and bound procedure then interprets this code list. The user inputs the objective and constraints by programming it in Fortran 90; see §5 below.

GlobSol's automatic differentiation was designed for small problems whose code lists are not difficult to evaluate. Although some removal of redundant operations is done, both the structure of GlobSol's code list and interpretive nature of its evaluation lead to less efficiency than possible by hand-coding functions and derivatives. For example, at present, the code list for the function and gradients is evaluated in its entirety, even if only a specific gradient component is required.

The operator overloading technique in GlobSol is similar to the "tape" concept in the ADOLC automatic differentiation system [8]. We give simple examples of our code list creation scheme in [16, §1.4 and §2.2.2 to §2.2.4].

An early reference on automatic differentiation is [28], while well-edited conference proceedings include [9, 1, 2]. A recent book detailing sophisticated efficiency measures is [7].

2.2 Constraint Propagation

GlobSol uses constraint propagation as an efficiency-gaining device in its overall branch and bound algorithm. GlobSol applies constraint propagation at the level of individual operations in the code list. For example, such an operation may be

$$x_p = x_q + x_r,$$

where x_p, x_q, and x_r are intermediate variables in the process of evaluating and objective, constraint, gradient, or Hessian matrix. If, say, better bounds on x_p are obtained, then a constraint propagation step would be to form $x_q = x_p - x_r$ and $x_r = x_p - x_q$ to try to obtain better bounds on x_q and x_r. This is in contrast to some systems, where constraint propagation is applied across larger expressions, or where the system itself is written in a language specifically designed for constraint propagation. We give examples of our use of constraint propagation in [16, Chapter 7], while our original study of the technique appears in [13].

A system more heavily emphasizing constraint propagation than GlobSol is Numerica [31]. However, the designers of Numerica have also recognized that constraint propagation alone could not provide all the needed efficiency in the branch and bound algorithm, when there is significant coupling between the equations of the gradient; Newton-type iterations are used then.

2.3 Interval Newton Methods

Interval Newton methods are perhaps the most important acceleration technique for larger problems. Abstractly, an interval Newton method is of the form

$$\tilde{x} = N(F; x, \check{x}), \tag{4}$$

where F represents a system of nonlinear equations as in (1), where x is a domain box, where $\check{x}$, generally taken so $\check{x} \in x$, is an initial guess point, and where $\tilde{x}$ is the image of the interval Newton method.

The primary advantages of an interval Newton method are the following.

1. Under conditions that usually hold for small enough box widths, interval Newton methods reduce the diameter (i.e. reduce the maximum coordinate width) of a box x at a rate such that the diameter $d(\tilde{x})$ of the image $\tilde{x}$ obeys $d(\tilde{x}) = \mathcal{O}\left(d(x)^2\right)$. This is particularly true when F represents an approximately linear or quadratic system, and interval dependencies have been eliminated symbolically Also, obtaining $\tilde{x}$ can be done in $\mathcal{O}\left(n^3\right)$ operations, when F represents a dense system of n equations in n unknowns.
2. Any solutions x^*, $F(x^*) = 0$ with $x^* \in x$ must have $x^* \in \tilde{x}$.
3. Under weak additional conditions, if $\tilde{x} \subset x$, then this proves that there is a unique $x^* \in x$ such that $F(x^*) = 0$.

Numerous authors have discussed interval Newton methods. We introduce them (with more pointers to the literature) in [16, §1.5 and §6.2], and in [19]. See [27] for a careful theoretical treatment.

Within GlobSol, our interval Newton method is used with the Fritz–John (i.e. generalized Lagrange multiplier) system to reduce the sizes of boxes x produced during the subdivision process. However, GlobSol does not presently use the Fritz–John system within the context of verifying that given boxes contain unique solutions; instead, GlobSol works with the system of equality constraints $c(x) = 0$ and active inequality constraints $g(x)$ to prove existence of a feasible point. This is because, even though some coordinates can be reduced when the Fritz–John system is used with the interval Gauss–Seidel method over large boxes, we have found practical difficulties getting interval Newton methods applied to the Fritz–John system to reduce each coordinate. (In some cases, the interval Newton method may only be effective over very small boxes, or good Lagrange multiplier estimates cannot be produced.)

GlobSol combines the interval Gauss–Seidel method with special preconditioners. The interval Gauss–Seidel method generally produces narrower image boxes $\tilde{x}$ than, say, the Krawczyk method. (See [27, p. 138].) Furthermore, our Linear programming preconditioners for the interval Gauss–Seidel method, described in [16, Chapter 3], are more effective for wide boxes x; also, our timings within GlobSol have shown that obtaining these preconditioners is not overly costly within the overall branch and bound algorithm.

2.4 Additional Techniques

Use of Already-Found Optima. One additional fairly common technique is use of approximate optima $\tilde{x}$ to eliminate portions of the original box x that do not contain optima other than those already found. We explain this in [17]. GlobSol contains a special variant of the technique that avoids excessive work when the Jacobi matrix of F is ill-conditioned or singular near solutions; see

[16, step 4(c), p. 148]. This is closely related to the procedure in [12, step 4, Algorithm 2.6].

Van Iwaarden refers to such use of approximate optima as "back-boxing" [32].

Bound Constraints. GlobSol has a special pre-processing step, called "peeling," used to handle bound constraints, when one or more of the lower coordinate bounds $\underline{x}_i$ or upper coordinate bounds $\overline{x}_i$ of the original box x represents not only a limit of the search region, but also a bound constraint. See [16, §5.2.3], and see §5 below for advice on use of this technique.

Choice of Coordinate to Bisect. GlobSol uses a scaled version of "maximum smear" advocated by Ratz [29]. (See also [16, (5.1), p. 175].) For constrained problems, the technique is applied to the sum of the objective function and active constraints.

3   History of GlobSol

Our original validated nonlinear systems solver, developed jointly with Manuel Novoa, was INTBIS, now an *ACM Transactions on Mathematical Software* algorithm [22]. In this small FORTRAN 77 package, polynomial systems can be solved with validation, merely by inputting the coefficients, in the format described in [26]. INTBIS follows closely [12, Algorithm 2.6], where the "root inclusion test" $\mathcal{T}_F$ of step 3 of that algorithm is represented by an interval Newton method. The interval Newton method in INTBIS differs from that of GlobSol mainly in the use of preconditioners: Jacobi matrices are preconditioned only by the inverse of the midpoint matrix, and not by the special linear programming preconditioners of [16, Ch. 3].

Subsequently, we provided a Fortran 77 interval transcendental function library INTLIB [20], with which users of INTBIS could write their own function and Jacobi matrix routines, for more or less arbitrary functions. Stadtherr et al have specially modified INTBIS to obtain impressive success in solving practical problems in chemical engineering, including an early success with larger sparse problems [30], parameter-fitting problems where previous non-verified solutions were erroneous [5], etc. Stadtherr continues to use these specially developed codes derived from INTBIS, as the overhead associated with using interpretive code lists, etc. in GlobSol appears to add substantially to execution time (perhaps a factor of 10 in some cases). The main disadvantage, besides certain outdated algorithms in the distribution version of INTBIS, is that objective and residuals must be programmed with subroutine calls, in an assembly-language-like way.

Prior to development of the automatic differentiation foundation of GlobSol, we experimented in FORTRAN 77 with our version of constraint propagation ("substitution-iteration", [13]) and with bound-constrained optimization [14]; in [14], we introduced both our aforementioned "peeling" idea and experimented with our special preconditioners as well.

During our experiments, we found ourselves struggling to hand-program various examples. This led us to initially develop the Fortran 90 interface to INTLIB (subsequently polished, with John Reid's help, and published as module INTERVAL_ARITHMETIC [15]). At that time, with aid from a National Science Foundation grant, we also developed the basic automatic differentiation interface and interpreters, represented in GlobSol by module CODELIST_CREATION in overload/overload.f90 and various interpreters in GlobSol's function subdirectory, such as FORWARD_SUBSTITUTION_POINT in the file function/intermediate_variable_evaluators/forwsubpt.f90.
This early version of our package, which we called "INTOPT90", contained a separate nonlinear systems solver and optimizer. It also contained various alternate specific subalgorithms (such as different preconditioner computations, interval Gauss–Seidel versus interval Gaussian elimination, and different normalizations for the Fritz–John equations), as well as a facility for performance statistics, including timing information. While much of the specific information about our package in [16] remains valid for the present version of GlobSol, some of that information applies only to this early version.

In 1998–1999, collaborating with George Corliss and others under G. William Walster in a Sun Microsystems Cooperative Research and Development project, we extensively polished the user interface to GlobSol, including providing scripts to unify the code list creation, code list optimization, and search phases, cleaning the configuration file, etc. At this time, we also enhanced the underlying package, providing inequality constraints, providing additional functions recognizable by the code list creator and interpreter, making certain changes to increase efficiency, etc. We also reorganized the directory structure in which GlobSol is shipped to make it clearer, and we reorganized the top-level subroutine calling structure to make it easier to call GlobSol and utilize its results in other packages. As part of this Sun Microsystems project, we did extensive formal testing, leading to the discovery and correction of significant numbers of bugs. The Sun Microsystems project also prominently included utilization of GlobSol for various practical problems [3].

Unfortunately, due to constraints on our time, not all of the possible experimental paths (e.g. not all possible formulations of the Fritz–John equations) have been uniformly tested; only those previously proven to be generally better were tested. Users of GlobSol should not find problems with default configurations; see §5 below.

We added additional capabilities to the distribution version of GlobSol subsequent to completion of the Sun Microsystems project. For example, we provided a facility for dealing with non-zero-width parameters, considered as "constants," but alterable after the code list is created; see [24] for a study of such parameters.

The distribution version of GlobSol is presently available at
http://interval.louisiana.edu/GlobSol/download_GlobSol.html
This includes source code and installation scripts, free of charge.

In addition to the distribution version of GlobSol, we maintain an experimental version, with capabilities that, due to either licensing restrictions or lack of thorough testing, are not presently included in the distribution version. See §6 below.

Any history of interval-based global optimization should mention Eldon Hansen, whose pioneering techniques are summarized in [10].

4   GlobSol's Overall Algorithm

Verified solution of a global optimization problem proceeds most conveniently with the "globsol" script, which does the following:

1. Compile and run the user-provided Fortran 90 program defining the objective and constraints, to produce the code list.
2. Optimize the code list (removing redundant operations).
3. Run the global search algorithm.
4. Clean the directory of temporary files.

The actual global search algorithm consists of

1. Initial I/O (in routine INITIALIZE_FIND_GLOBAL_MIN in file f90intbi/initialize_find_global_min.f90)
2. The "peeling" process of [16, §5.2.3] (in routine PROCESS_INITIAL_BOX in file f90intbi/process_initial_box.f90.
3. The actual global search routine (in routine RIGOROUS_GLOBAL_SEARCH in file f90intbi/rigorous_global_search.f90)
4. Final I/O (in routine FINISH_FIND_GLOBAL_MIN in file f90intbi/finish_find_global_min.f90)

These four routines are called from the driver routine FIND_GLOBAL_MIN in file f90intbi/find_global_min.f90.

The actual search routine RIGOROUS_GLOBAL_SEARCH contains the following algorithm.

Algorithm 1 *(GlobSol's global search algorithm)*
INPUT: A list $\mathcal{L}$ of boxes x to be searched.
OUTPUT: A list $\mathcal{U}$ of small boxes and a list $\mathcal{C}$ of boxes verified to contain feasible points, such that any global mimimizer must lie in a box in $\mathcal{U}$ or $\mathcal{C}$.
DO WHILE *($\mathcal{L}$ is non-empty)*

1. *Remove a box x from the list $\mathcal{L}$.*
2. *IF x is sufficiently small THEN*
    (a) Place x on either $\mathcal{U}$ or $\mathcal{C}$.
    (b) CYCLE
    END IF
3. *(Constraint Propagation)*
    (a) Use constraint propagation to possibly narrow the coordinate widths of the box x.
    (b) IF constraint propagation has shown that x cannot contain solutions THEN CYCLE

4. *(Interval Newton)*
   (a) *Perform an interval Newton method to possibly narrow the coordinate widths of the box **x**.*
   (b) IF *the interval Newton method has shown that **x** cannot contain solutions* THEN CYCLE
5. IF *the coordinate widths of **x** are now sufficiently narrow* THEN
   (a) *Place **x** on either* $\mathcal{U}$ *or* $\mathcal{C}$.
   (b) CYCLE
6. *(Subdivide)*
   (a) *Choose a coordinate index k to bisect (i.e. to replace $[\underline{x}_k, \overline{x}_k]$ by $[\underline{x}_k, (\underline{x}_k + \overline{x}_k)/2]$ and $[(\underline{x}_k + \overline{x}_k)/2, \overline{x}_k]$).*
   (b) *Bisect **x** along its k-th coordinate, forming two new boxes; place these boxes on the list $\mathcal{L}$.*
   (c) CYCLE

END DO

END ALGORITHM 1
Notes:

1. Traditional techniques, such as the "midpoint test" and "gradient test" for rejecting boxes with high values of the global optimum or boxes that cannot contain critical points, are included in steps 3 and 4, but are irrelevant here.
2. Determining when a box **x** is "small enough" is more sophisticated than simply checking the widths of the coordinates; see [24] for details.
3. Steps 2 and 5 of Algorithm 1 are more involved in GlobSol than simply placing the result box on the list. In particular, an attempt is made to find approximate optima and do ϵ-inflation (as in [16, §4.2]), to avoid excessive subdivisions around solutions. The process includes the box complementation scheme of [16, §4.3.1]. Embodied in internal subroutine HANDLE_LEAF in file f90intbi/rigorous_global_search.f90. This process is subject to change in the future, with an eye towards simplification.

GlobSol has numerous sub-algorithms. For example, when there are equality constraints, the "midpoint test" for determining a rigorous upper bound on global minima necessarily is more complicated than evaluating the objective function at a point. In particular, the point of evaluation must be known to be feasible. We explain the process in GlobSol for verifying feasibility in [18] and [16, §5.2.4]. This process is presently a weakness that prevents some equality-constrained problems from being handled as efficiently as unconstrained or certain inequality-constrained problems.

GlobSol also attempts to find approximate feasible points, for use in the midpoint test. A generalized Newton method is used for this purpose. See the report "An Iterative Method for Finding Approximate Feasible Points" at

http://interval.louisiana.edu/GlobSol/
Dian-approximate-optimizer.pdf

5   Use of GlobSol: Examples and Advice

Once installed, GlobSol requires the following files to run.

GlobSol.CFG: The GlobSol configuration file, this comes with default settings
    that the user need not initially change; however, see below.
*.DT?: The box data file, this user-supplied file defines the limits of the search
    box x and specifies which of the coordinate bounds are to be considered
    bound constraints for the purposes of "peeling".
*.f90: The user supplies a Fortran 90 program to define the objective and
    constraints.

An example is the simple illustration of mixed constraints, found in the
integration_test_data subdirectory of GlobSol. The supplied box data file,
named "mixed.DT1", is:

```
1D-5
0   1
0   1
F F
F F
```

The first line signifies a tolerance of 10^{-5}; actual answer boxes can be expected to have relative widths up to the square root of this tolerance. The second
and third lines specify bounds $x_1 = [0, 1]$ and $x_2 = [0, 1]$ on the initial search
region x. The last two lines specify that none of the search region bounds are to
be considered as bound constraints for the purposes of the peeling process.

The supplied Fortran 90 source file, named "mixed.f90", is:

```
PROGRAM SIMPLE_MIXED_CONSTRAINTS
 USE CODELIST_CREATION

     PARAMETER (NN=2)
     TYPE(CDLVAR), DIMENSION(NN):: X
     TYPE(CDLLHS), DIMENSION(1):: PHI
     TYPE(CDLINEQ), DIMENSION(2):: G
     TYPE(CDLEQ), DIMENSION(1) :: C

     CALL INITIALIZE_CODELIST(X)

 PHI(1) = -2*X(1)**2 - X(2)**2
 G(1) = X(1)**2 + X(2)**2 - 1
 G(2) = X(1)**2 - X(2)
 C(1) = X(1)**2 - X(2)**2

     CALL FINISH_CODELIST
END PROGRAM SIMPLE_MIXED_CONSTRAINTS
```

Here, the objective is defined with the special variable type CDLLHS ("code
list left hand side"), the inequality constraints with CDLINEQ, and the equality constraints with CDLEQ. (Observe the slight difference with the explanation
in [16].) Each of these variables should be considered "write-once;" that is, they
should appear only once and in left-hand-sides of assignment statements.

Issuing the command "globsol mixed 1" invokes the GlobSol script to produce the output file mixed.OT1. Here is an abridgement of this file:

```
Output from FIND_GLOBAL_MIN on  07/28/2002  at  14:53:47.
Version for the system is: October 10, 2000

Codelist file name is: mixedG.CDL
Box data file name is: mixed.DT1

Initial box:
[    0.0000D+00,    0.1000D+01 ] [    0.0000D+00,    0.1000D+01 ]
BOUND_CONSTRAINT:
  F F   F F
----------------------------------------
CONFIGURATION VALUES:
EPS_DOMAIN:   0.1000D-04   MAXITR:    20000
MAX_CPU_SECONDS:    0.3600D+04
DO_INTERVAL_NEWTON: T  QUADRATIC: T  FULL_SPACE: F
...
(additional configuration variables are printed here.)
...
Default point optimizer was used.

THERE WERE NO BOXES IN THE LIST OF SMALL BOXES.
LIST OF BOXES CONTAINING VERIFIED FEASIBLE POINTS:
Box no.:            1
Box coordinates:
[    0.7071D+00,    0.7071D+00 ] [    0.7071D+00,    0.7071D+00 ]
PHI:
[   -0.1500D+01,   -0.1500D+01 ]
B%LIUI(1,*):
F F
B%LIUI(2,*):
F F
B%SIDE(*):
F F
B%PEEL(*):
T T
Level:              1
Box contains the following approximate root:
   0.7071D+00    0.7071D+00
OBJECTIVE ENCLOSURE AT APPROXIMATE ROOT:
[   -0.1500D+01,   -0.1500D+01 ]
Unknown = T    Contains_root = T
Changed coordinates:
T F
UO:
[    0.3852D+00,    0.3852D+00 ]
U:
[    0.5777D+00,    0.5777D+00 ] [    0.0000D+00,    0.1000D+01 ]
V:
[    0.1926D+00,    0.1926D+00 ]
INEQ_CERT_FEASIBLE:
  F T
NIN_POSS_BINDING:            1
----------------------------------------------------
ALGORITHM COMPLETED WITH LESS THAN THE MAXIMUM NUMBER,
      20000  OF BOXES.
Number of bisections:            1
No. dense interval residual evaluations -- gradient code list:         53
...
(Additional performance information is printed here.)
...
Total number of boxes processed in loop:         4
BEST_ESTIMATE:   -0.1500D+01
Overall CPU time:    0.1001D-01
CPU time in PEEL_BOUNDARY:    0.0000D+00
CPU time in REDUCED_INTERVAL_NEWTON:    0.0000D+00
```

Here, the "default point optimizer" is the generalized Newton method of [21]. The components LIUI, SIDE, and PEEL of the box data type B need not concern the user. Generally, the "approximate root" is the midpoint of the box, while UO, U, and V represent bounds on the generalized Lagrange multipliers for the objective, inequality constraints, and equality constraints. Bounds of $[0, 1]$ often correspond to inequality constraints that are not active; this is the case for this problem, since, from the line below the Lagrange multiplier bounds, we see that the second inequality constraint is "certainly feasible," meaning an interval evaluation gave a non-positive result[2] . The BEST_ESTIMATE is the best estimate for the global minimum, obtained through point evaluations or, when equality constraints are present, through evaluation of the objective over a small box within which a feasible point has been proven to exist. (This value is called $\overline{f}$ in some of the literature.)

5.1   Advice on Interpretation of Results

The only guarantees with GlobSol's algorithm are that any possible global minimizers must be contained in the union of the two lists of output boxes. Occasionally, the problem will not have global minimizers, but one or both lists are non-empty[3] . If a reasonable BEST_ESTIMATE has been found, then the boxes in the output lists should have relatively low objective values, even if they don't correspond to minimizers.

Conversely, GlobSol may complete with both the list of small boxes and list of boxes containing verified feasible points empty. For example, we expect both lists to be empty for a linear objective function without constraints. Adding constraints (or setting a sufficient number of flags in the box data file to indicate bound constraints) will cause GlobSol to output non-empty lists. Conceptually (although not from an implementation point of view), one can think of box limits not corresponding to bound constraints as topologically "open," and box limits corresponding to bound constraints as "closed"; extrema are guaranteed to exist only over closed, bounded regions.

5.2   Advice on the Configuration File

With numerous in-line comments, GlobSol's configuration file is to a large extent self-documenting. The file is organized into sections, consisting of:

1. limits associated with code list creation,
2. switches to control printing,
3. stopping tolerances,
4. limits on the algorithm, such as CPU time,
5. miscellaneous algorithm controls.

---

[2] If a constraint is certainly feasible over a box, the constraint is dropped from the computation, but the corresponding Lagrange multiplier is actually zero. A future version of GlobSol will print "0" for such Lagrange multipliers.

[3] This can happen, for example, for a monotonic function without constraints, where interval overestimation prevents monotonicity from being detected.

The limits associated with code list creation are necessary since not all memory could be dynamically allocated during this process. GlobSol should give an appropriate error message indicating which of these limits has been exceeded, if it is necessary to change these.

Levels of printing are independently available for the overall algorithm and various sub-algorithms. Most of these levels are for debugging, and will result in excessive output on practical problems. However, some may reveal useful information to aid in solving problems efficiently. Users may wish, for example, to set `PRINTING_IN_OVERALL_ALGORITHM` to 2 (from its default of 0).

Resource Limits and Tolerances. We recommend setting `MAX_CPU_SECONDS` as desired, and setting `MAXITR`, representing the maximum number of boxes to be processed, higher if the default is exceeded.

Stopping tolerances may possibly be changed from defaults to enable GlobSol to complete for certain problems. Users should read [24] to understand the meaning of these tolerances.

Except as noted below, we do not envision it as usually appropriate for the user to change other switches that control the algorithms.

Turning On and Off Constraint Propagation. Constraint propagation as explained in §2.2 above is "on" by default; experimentation has shown that, within GlobSol's overall algorithm, this constraint propagation usually saves significant amounts of computational effort. However, constraint propagation is not useful for a few problems. Furthermore, since GlobSol does not yet have full implementations for all inverses of all functions supported by the code list, constraint propagation is not possible for certain functions. GlobSol's constraint propagation at the code list level can be turned off by setting the configuration variable `USE_SUBSIT` to "F".

We at one time contemplated having the user define relations and inverses at a higher level than the elemental operations of the code list, for use in constraint propagation. Although this may be useful, it is not fully implemented in the present version of GlobSol.

5.3   Advice on Using the "Peeling" Process

The peeling process is an effective way of handling bound constraints when the number of variables n is small or when the number of actual bound constraints is small, especially when the number of additional equality constraints is small. This is because, during the process, we evaluate the objective on lower-dimensional sub-boxes representing the bound constraints and intersections of bound constraints, and can thus obtain better upper bounds (i.e. better `BEST_ESTIMATE` or $\overline{f}$) with minimal overestimation due to interval dependency.

However, although only subfaces with sufficiently low `BEST_ESTIMATE` are processed, peeling generates up to 2^m sub-boxes, where m is the number of bound constraints that have been set. (The number is 3^m if both lower and upper bounds correspond to bound constraints.) Therefore, for large numbers of variables, it is not advisable to set large numbers of bound constraints for peeling. One alternative is to selectively set a few bound constraints, experimenting until GlobSol gives non-empty answer lists. Another alternative is to make the search box slightly larger, and define the bound constraints as equality constraints. However, see below.

5.4 Advice on Equality Constraints

A present weakness in GlobSol is its treatment of equality constraints. In particular, because GlobSol verifies feasibility by forming a system from the equality and active inequality constraints, the number of equality constraints cannot exceed the number of unknowns. We are presently working on alternate paradigms to circumvent this problem. (See below.)

5.5 Advice on Data Fitting

GlobSol presently does not seem to handle data fitting problems (least squares, minimax, or least absolute value) well when there are large numbers of data points. (However, see [5].) We are presently working on an alternate paradigm; see §6.1 below.

6 The Experimental Version and GlobSol's Future

Nonlinear Systems. We have quit maintaining the separate nonlinear system solver that was in early versions of the GlobSol package. Instead, in our experimental version of GlobSol, we are including a "nonlinear system" algorithm path. This allows us to maintain a single overall algorithm, as well as to take advantage of GlobSol's inequality constraint handling capabilities.

Taylor Models. Our experimental version of GlobSol includes an ability to switch on and off Taylor arithmetic in evaluation of the objective and constraints, as well as to "symbolically" precondition interval Jacobi matrices with Taylor arithmetic, for interval Gauss–Seidel iteration;. See [2, 23] for results of using such Taylor models within our experimental version of GlobSol. For the Taylor arithmetic, we have interfaced GlobSol with the COSY-Infinity package of Berz et al, using Jens Hoefkens' Fortran 90 module [11].

6.1 Nonlinear Data Fitting

Although least squares, minimax, and ℓ_1 data fitting problems can be formulated within the present distribution GlobSol environment using Lemaréchal's conditions [25], directly as sums of squares, or with use of the augmented system (using the LEAST_SQUARES configuration variable), these approaches lead to excessive computation times for many problems. We speculate that this is because, with many data points, the resulting constraints or equations are approximately linearly dependent.

In [33], we proposed a new paradigm for linear least squares. This paradigm can be adapted to the nonlinear case. We have an algorithm path in the experimental version of GlobSol within which we are presently experimenting with this possibility.

References

[1] M. Berz, C. Bischof, G. Corliss, and A. Griewank, editors. *Computational Differentiation: Techniques, Applications, and Tools*, Philadelphia, 1996. SIAM.

[2] G. Corliss, Ch. Faure, A. Griewank, L. Hascoët, and U. Naumann, editors. *Automatic Differentiation of Algorithms: From Simulation to Optimization*, New York, 2002. Springer-Verlag.

[3] G. F. Corliss and R. B. Kearfott. Rigorous global search: Industrial applications. In *Developments in Reliable Computing*, pages 1–16, Dordrecht, Netherlands, 2000. Kluwer.

[4] J. E. Dennis and R. B. Schnabel. *Numerical Methods for Unconstrained Optimization and Nonlinear Least Squares*. Prentice-Hall, Englewood Cliffs, NJ, 1983.

[5] C.-Y. Gau and M. A. Stadtherr. Nonlinear parameter estimation using interval analysis. *AIChE Symp. Ser*, 94(304):445–450, 1999.

[6] P. E. Gill, W. Murray, and M. Wright. *Practical Optimization*. Academic Press, New York, 1981.

[7] A. Griewank. *Evaluating Derivatives: Principles and Techniques of Algorithmic Differentiation*. Frontiers in Applied Mathematics. SIAM, Philadelphia, 2000.

[8] A. Griewank. ADOL-C, a package for automatic differentiation of algorithms written in C/C++, 2002. http://www.math.tu-dresden.de/wir/project/adolc/.

[9] A. Griewank and ed. Corliss, G. F., editors. *Automatic Differentiation of Algorithms: Theory, Implementation, and Application*, Philadelphia, 1991. SIAM.

[10] E. R. Hansen. *Global Optimization Using Interval Analysis*. Marcel Dekker, Inc., New York, 1992.

[11] J. Hoefkens. *Rigorous Numerical Analysis with High-Order Taylor Models*. PhD thesis, Department of Mathematics, Michigan State University, 2001.

[12] R. B. Kearfott. Abstract generalized bisection and a cost bound. *Math. Comp.*, 49(179):187–202, July 1987.

[13] R. B. Kearfott. Decomposition of arithmetic expressions to improve the behavior of interval iteration for nonlinear systems. *Computing*, 47(2):169–191, 1991.

[14] R. B. Kearfott. An interval branch and bound algorithm for bound constrained optimization problems. *Journal of Global Optimization*, 2:259–280, 1992.

[15] R. B. Kearfott. Algorithm 763: INTERVAL_ARITHMETIC: A Fortran 90 module for an interval data type. *ACM Trans. Math. Software*, 22(4):385–392, December 1996.

[16] R. B. Kearfott. *Rigorous Global Search: Continuous Problems*. Kluwer, Dordrecht, Netherlands, 1996.

[17] R. B. Kearfott. Empirical evaluation of innovations in interval branch and bound algorithms for nonlinear algebraic systems. *SIAM J. Sci. Comput.*, 18(2):574–594, March 1997.

[18] R. B. Kearfott. On proving existence of feasible points in equality constrained optimization problems. *Math. Prog.*, 83(1):89–100, September 1998.

[19] R. B. Kearfott. Interval analysis: Interval Newton methods. In *Encyclopedia of Optimization*, volume 3, pages 76–78. Kluwer, 2001.

[20] R. B. Kearfott, M. Dawande, K.-S. Du, and C.-Y. Hu. Algorithm 737: INTLIB, a portable FORTRAN 77 interval standard function library. *ACM Trans. Math. Software*, 20(4):447–459, December 1994.

[21] R. B. Kearfott and J. Dian. An iterative method for finding approximate feasible points, 1998. preprint, http://interval.louisiana.edu/GlobSol/Dian-approximate-optimizer.pdf.

[22] R. B. Kearfott and M. Novoa. Algorithm 681: INTBIS, a portable interval Newton/bisection package. *ACM Trans. Math. Software*, 16(2):152–157, June 1990.

[23] R. B. Kearfott and Walster G. W. Symbolic preconditioning with Taylor models: Some examples, 2001. accepted for publication in *Reliable Computing*.

[24] R. B. Kearfott and G. W. Walster. On stopping criteria in verified nonlinear systems or optimization algorithms. *ACM Trans. Math. Software*, 26(3):373–389, September 2000.

[25] C. Lemaréchal. Nondifferentiable optimization. In M. J. D. Powell, editor, *Nonlinear Optimization 1981*, pages 85–89, New York, 1982. Academic Press.

[26] A. P. Morgan. *Solving Polynomial Systems Using Continuation for Engineering and Scie ntific Problems*. Prentice-Hall, Englewood Cliffs, NJ, 1987.

[27] A. Neumaier. *Interval Methods for Systems of Equations*. Cambridge University Press, Cambridge, England, 1990.

[28] L. B. Rall. *Automatic Differentiation: Techniques and Applications*. Lecture Notes in Computer Science no. 120. Springer, Berlin, New York, etc., 1981.

[29] D. Ratz and T. Csendes. On the selection of subdivision directions in interval branch-and-bound methods for global optimization. *J. Global Optim.*, 7:183–207, 1995.

[30] C. A. Schnepper. *Large Grained Parallelism in Equation-Based Flowsheeting Using Interval Newton / Generalized Bisection Techniques*. PhD thesis, University of Illinois, Urbana, 1992.

[31] P. Van Hentenryck, L. Michel, and Y. Deville. *Numerica: A Modeling Language for Global Optimization*. MIT Press, Cambridge, MA, 1997.

[32] R. J. Van Iwaarden. *An Improved Unconstrained Global Optimization Algorithm*. PhD thesis, University of Colorado at Denver, 1996.

[33] J. Yang and R. B. Kearfott. Interval linear and nonlinear regression: New paradigms, implementations, and experiments, or new ways of thinking about data fitting, 2002, available at http://interval.louisiana.edu/preprints/2002_SIAM_minisymposium.pdf.

LaGO – An Object Oriented Library for Solving MINLPs[*]

Ivo Nowak[**], Hernán Alperin, and Stefan Vigerske

Humboldt-Universität zu Berlin, Institut für Mathematik
Rudower Chaussee 25, D-12489 Berlin, Germany
`ivo@mathematik.hu-berlin.de`
`http://www-iam.mathematik.hu-berlin.de/~eopt/`

Abstract. The paper describes a software package called LaGO for solving nonconvex mixed integer nonlinear programs (MINLPs). The main component of LaGO is a convex relaxation which is used for generating solution candidates and computing lower bounds of the optimal value. The relaxation is generated by reformulating the given MINLP as a block-separable problem, and replacing nonconvex functions by convex underestimators. Results on medium size MINLPs are presented.

Keywords: Mixed integer nonlinear programming, convex relaxation, heuristics, decomposition, software

AMS classifications: 90C22, 90C20, 90C27, 90C26, 90C59

1 Introduction

Several strategies for solving nonconvex MINLPs have been proposed [BP03]. Exact solution approaches include branch-and-reduce [TS02], branch-and-bound [ADFN98, SP99], interval analysis [VEH96] and outer approximation [KAGB01]. On the other side, heuristic solution algorithms explicitly addressing MINLP, such as the multistart scatter search heuristic [ULP+02], appeared rather rarely in the literature. More information on global optimization algorithms can be found in [Flo00, HP95, BSV+01].

LaGO (**La**grangian**G**lobal**O**ptimizer) is an objected oriented library for solving nonconvex mixed integer nonlinear programs (MINLPs) written in C++. The basic component of the solver is a relaxation, which is used for generating solution candidates and computing lower bounds of the optimal value. It is very important that the quality of the relaxation is good enough, in the sense that the amount of work to retrieve a good solution from the relaxation is not too high. For improving a given relaxation, a series of operations, which can be performed relatively fast and take advantage of a partially separable structure of the given optimization problem, is used.

---

[*] The work was supported by the German Research Foundation (DFG) under grant NO 421/2-1.
[**] corresponding author

C. Bliek et al. (Eds.): COCOS 2002, LNCS 2861, pp. 32–42, 2003.

The software can be used as a general purpose MINLP solver or as a toolbox for developing specialized solution algorithms. The object oriented design of the library allows easy replacement of modules, such as linear algebra routines and various solvers. Optimization problems can be either provided in AMPL-format [FGK93] or coded in C++ using LaGO 's class `MinlpProblem`. The solver supports two types of structural properties: *partial separability*, i.e. the Hessians have almost block-diagonal structure, and *sparse and low rank Hessians*, i.e. there exist fast methods for multiplying a vector with a Hessian.

The basic components of the current version of LaGO are: *preprocessing*, *convex relaxation* and *relaxation-based heuristic*, which are described in Sections 2, 3 and 4. A *branch-and-cut algorithm* is currently under development and will be added in the future. In Section 5 we report preliminary results, and give conclusions in Section 6.

Notation. Let $J \subset \{1, \ldots, n\}$ be an index set. We denote by $|J|$ the number of elements of J. A subvector $x_J \in \mathbb{R}^{|J|}$ of $x \in \mathbb{R}^n$ is defined by $(x_j)_{j \in J}$. Similarly, a function $f_J(x)$ is defined by $(f_j(x))_{j \in J}$. The space of k times differentiable functions $f : \mathbb{R}^n \mapsto \mathbb{R}^m$ is denoted by $C^k(\mathbb{R}^n, \mathbb{R}^m)$.

2 Preprocessing

We assume that the given MINLP problem has the form:

$$
\text{(P)} \qquad
\begin{aligned}
\min \quad & h_0(x) \\
\text{s.t.} \quad & h_I(x) \leq 0 \\
& h_E(x) = 0 \\
& x \in Y
\end{aligned}
$$

where

$$
Y = \{x \in [\underline{x}, \overline{x}] \mid x_j \in \{\underline{x}_j, \overline{x}_j\} \text{ for } j \in B\},
$$

$\underline{x}, \overline{x} \in \mathbb{R}^n$, $B \subseteq \{1, \ldots, n\}$, $I \cup E = \{1, \ldots, m\}$ with $I \cap E = \emptyset$, and $h_i \in C^2(\mathbb{R}^n, \mathbb{R})$ for $i = 0, \ldots, m$. Note that MINLPs with piecewise twice differentiable functions and integrality constraints, $x_i \in [\underline{x}_i, \overline{x}_i] \cap \mathbb{Z}$, can be transformed into the form (P) by introducing additional binary variables.

To facilitate the construction of a relaxation, problem (P) is transformed into the following *block-separable* extended reformulation.

$$
\text{(P}_{\text{ext}}\text{)} \qquad
\begin{aligned}
\min \quad & c^T x + c_0 \\
\text{s.t.} \quad & A_I x + b_I \leq 0 \\
& A_E x + b_E = 0 \\
& g^k(x_{J_k}) \leq 0, \qquad k = 1, \ldots, p \\
& x \in Y
\end{aligned}
$$

where $\{J_1, \ldots, J_p\}$ is a partition of $\{1, \ldots, n\}$, i.e. $\bigcup_{k=1}^p J_k = \{1, \ldots, n\}$ and $J_i \cap J_k = \emptyset$ for $i \neq k$, $g^k \in C^2(\mathbb{R}^{n_k}, \mathbb{R}^{m_k})$, $k = 1, \ldots, p$, with $n_k = |J_k|$, $c \in \mathbb{R}^n$,

$b \in \mathbb{R}^m$, $A_I \in \mathbb{R}^{(|I|,n)}$ and $A_E \in \mathbb{R}^{(|E|,n)}$. To obtain the block functions $g^k(x_{J_k})$ used in (P_{ext}) we construct the *sparsity graph* $G_s = (V, E_s)$ with the nodes $V = \{1, \ldots, n\}$ and the edges

$$E_s = \left\{ (i,j) \in V^2 \ \middle| \ \frac{\partial^2 h_k(x)}{\partial x_i \partial x_j} \neq 0 \text{ for some } k \in \{0, \ldots, m\} \text{ and } x \in [\underline{x}, \overline{x}] \right\}.$$

Let $\{\tilde{J}_1, \ldots, \tilde{J}_p\}$ be a partition of V. We define the set of nodes of $\bigcup_{l=k+1}^{p} \tilde{J}_l$ connected to $\tilde{J}_k$ by $R_k = \{i \in \bigcup_{l=k+1}^{p} \tilde{J}_l \mid (i,j) \in E_s, j \in J_k\}$, for $k = 1, \ldots, p$. The set R_k can be interpreted as the set of flows of a nonlinear network problem connecting a component $\tilde{J}_k$ with components $\tilde{J}_l$, where $k < l$. After introducing new variables $y^k \in \mathbb{R}^{|R_k|}$ for every set R_k, and using the copy-constraints $x_{R_k} = y^k$, problem (P) is formulated as a block-separable program with respect to the blocks $J_k = (\tilde{J}_k, R_k)$, with $k = 1, \ldots, p$. Finally, nonlinear equality constraints $h_i(x) = 0$ for $i \in E$ are replaced by the two inequality constraints $h_i(x) \leq 0$ and $-h_i(x) \leq 0$, and block-separable constraints

$$\sum_{k=1}^{p} h_i^k(x_{J_k}) \leq 0$$

are replaced by

$$\sum_{k=1}^{p} t_{ik} \leq 0, \quad h_i^k(x_{J_k}) \leq t_{ik}, \quad k = 1, \ldots, p.$$

The resulting program has the form (P_{ext}), with $g_i^k(x_{J_k}, t_{ik}) = h_i^k(x_{J_k}) - t_{ik}$. For the sake of simplicity we keep the notation n, m, B, I, E, Y as in (P) including the new variables t_{ik} into the block x_{J_k}.

Furthermore, the type (linear, convex, concave) of all functions is determined by evaluating the minimum and maximum eigenvalue of each Hessian at sample points. Since the given MINLP is coded in AMPL [FGK93], all functions are given in a black-box representation, i.e. there exist only procedures for evaluating functions, gradients and Hessians. Therefore, the generation of the sparsity graph and the determination of function types is performed by sampling techniques.

Problem (P) assumes the existence of bounds $\underline{x}_j$ and $\overline{x}_j$ for all $j = 1, \ldots, n$. This is not the usual case in the instances collected from MINLPLib [BDM03] for instance. Since we need the box $[\underline{x}, \overline{x}]$ to compute convex relaxations and for the above sampling technique, we generate missing bounds by determining the maximum and minimum value of a variable over a domain defined by the convex constraints of problem (P). If there are not enough convex constraints to determine a variable bound, we use a simple guessing algorithm.

3   Convex Relaxation

A *convex relaxation* of problem (P_{ext}) is defined by

(C)
$$\begin{aligned}
\min\; & c^T x + c_0 \\
\text{s.t.}\; & A_I x + b_I \leq 0 \\
& A_E x + b_E = 0 \\
& \check{g}^k(x_{J_k}) \leq 0, \qquad k = 1,\dots,p \\
& x \in [\underline{x}, \overline{x}]
\end{aligned}$$

where $\check{g}^k$ are *convex underestimators* of g^k over $X_k = [\underline{x}_{J_k}, \overline{x}_{J_k}]$, i.e. $\check{g}^k(x) \leq g^k(x)$ for all $x \in X_k$ and $\check{g}^k$ is convex over X_k. The best convex underestimators are convex envelopes. Since computing convex envelopes of nonconvex functions can be computationally very expensive, we use the following procedure for generating convex relaxations.

We construct at first a *polynomial relaxation* of (P_{ext}) by replacing nonquadratic functions by *polynomial underestimators*. We assume that the nonquadratic functions of (P_{ext}) are of the form $g(x) = b^T x + f(x_N)$ where the number $|N|$ of nonlinear variables is small. We use as an underestimator for f a multivariate polynomial defined by

$$p(x) = a^T \varphi(x) \tag{1}$$

where $\varphi(x) = (x^{\beta_1}, \dots, x^{\beta_r})^T$, $\beta_j \in I\!N_0^n$, for $1 \leq j \leq r$, is a vector of monomials, and $a \in I\!R^r$ is the vector of coefficients for each monomial. The degree of the polynomial p is the number $d = \max_{j=1}^r |\beta_j|$ with $|\beta| = \sum_{i=1}^n \beta_i$. For our published results, we use $d = 2$, but in our implementation larger degree polynomial can be used. Let D^k be a differential operator defined by

$$D^k(f(x)) = \left(\frac{\partial^k f(x)}{\partial x^{\beta_l}} \right)_{1 \leq l \leq r_k},$$

with $|\beta_l| = k$ for $l = 1,\dots,r_k \leq \binom{r}{k}$ can be at most all the possible monomials of degree k, but in fact the sparsity pattern is used and fewer monomial than $\binom{r}{k}$ are needed. In order to determine the coefficients $a \in I\!R^r$ of the polynomial underestimator (1), we solve the linear program

(U)
$$\begin{aligned}
\min_{a \in I\!R^r}\; & \sum_{k=0}^2 \delta_k \sum_{x \in S_k} \|D^k(f(x) - a^T \varphi(x))\|_1 \\
\text{s.t.}\; & a^T \varphi(x) \leq f(x),\; x \in S_0.
\end{aligned}$$

The coefficients $\delta_0 > \delta_1, \delta_2$ give the relative importance to the information that comes from the evaluation of the function, the gradient and the Hessian respectively. The finite sample sets $S_k, k = 0, 1, 2$ contain sample points for the computation of the function, gradient and Hessian respectively. In general, this approach is only rigorous for certain kind of functions such as convex and concave.

The polynomial relaxation of (P_{ext}) can be reformulated by adding some extra variables and constraints as the following *mixed-integer quadratic program* (MIQQP)

$$\text{(Q)} \quad \begin{array}{ll} \min & c^T x + c_0 \\ \text{s.t.} & A_I x + b_I \leq 0 \\ & A_E x + b_E = 0 \\ & q^k(x_{J_k}) \leq 0, \qquad k = 1, \ldots, p \\ & x \in Y \end{array}$$

where q^k, $k = 1, \ldots, p$, are quadratic forms. We use two methods for convexifying (Q).

The first method is based on replacing all nonconvex quadratic forms by the so-called α-underestimators introduced by Adjiman and Floudas [AF97]. An α-underestimator of a function $f \in C^2(\mathbb{R}^n, \mathbb{R})$ is the function

$$\check{f}(x) = f(x) + \alpha^T r(x)$$

where

$$r(x) = \text{Diag}(x - \underline{x})(x - \overline{x}) \tag{2}$$

and $\text{Diag}(\cdot)$ denotes a diagonal matrix. The parameter $\alpha \in \mathbb{R}^n$ is computed according to

$$\alpha = \frac{1}{2} \max\{0, -\rho\} \, \text{Diag}(w)^{-2} e$$

where $e \in \mathbb{R}^n$ is the vector of ones, $w = \overline{x} - \underline{x}$ is the diameter vector of the interval, and

$$\rho \leq \lambda_1(\text{Diag}(w) \nabla^2 f(x) \, \text{Diag}(w)), x \in [\underline{x}, \overline{x}],$$

is a bound on the the minimum eigenvalue of the transformed Hessian of f over $[\underline{x}, \overline{x}]$. If f is a quadratic form, i.e. $\nabla^2 f$ is constant, ρ can be determined by eigenvalue computation. In the general case it can be computed by interval Hessians [AF97] or using a sampling technique. It is clear that $\check{f}(x) \leq f(x)$ for all $x \in [\underline{x}, \overline{x}]$, and $\check{f}$ is convex. Note that $f = \check{f}$ if f is convex.

Applying the α-underestimator technique directly to the original functions would also give a convex relaxation. However, our method is often tighter because the α-convexification depends only on the curvature of the function and not on the function behaviour. For more clarification see the example in Figure 1 where f is the original function, $\check{f}$ the α-convexification of f, q the polynomial underestimator, and $\check{q}$ the α-convexification of q.

The second method for convexifying problem (Q) is based on replacing the box-constraints $x_C \in [\underline{x}_C, \overline{x}_C]$ and the binary constraints $x_B \in \{\underline{x}_B, \overline{x}_B\}$ by the quadratic constraints $r_C(x) \leq 0$, and $r_B(x) = 0$ respectively, where the quadratic form r is defined as above, and $C = \{1, \ldots, n\} \setminus B$ is the index set of continuous variables. The resulting reformulation of (Q) is all-quadratic, and its

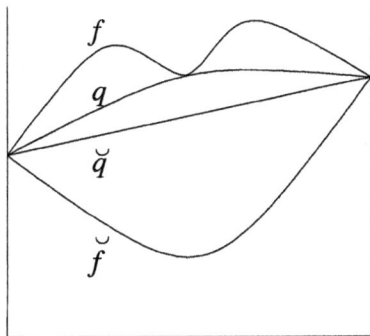

Fig. 1. α-convex underestimator versus the convexification of the polynomial underestimator

dual is a *semidefinite program*. There exist several methods for solving semidefinite programs [WSV00]. We use a method based on formulating the dual as an eigenvalue optimization problem [Now02]. This approach allows fast computation of near optimal dual solutions and supports decomposition. For each dual point μ produced by this algorithm, the Lagrangian is convex, and defines the following convex *Lagrangian relaxation*

$$(\text{R}_\mu) \qquad\qquad \min_{x \in [\underline{x}, \overline{x}]} L(x; \mu),$$

where $L(\cdot; \mu)$ is the Lagrangian to the all-quadratic programming reformulation of (Q).

Besides of the above convexification methods, one further method for constructing a convex relaxation (C) is implemented. It uses quadratic *convex global underestimators* proposed by Phillips, Rosen and Walke [PRW95]. Their method for approximating the convex envelope of a nonlinear function f by a quadratic function was improved in LAGO to reduce the absolute value of the smallest eigenvalue of the obtained quadratic underestimator. Let $S \subset [\underline{x}, \overline{x}]$ be a finite sample set, and define the quadratic function

$$q(x; a, b, c) = c + 2b^T x + x^T \operatorname{Diag}(a)x,$$

where $a, b \in I\!\!R^n$ and $c \in I\!\!R$. Then $q(\cdot; a, b, c)$ is convex, if and only if $a \geq 0$. The tightest quadratic convex underestimator $q(\cdot; a, b, c)$ over the set S is provided by the program

$$(\text{CGU}) \qquad \begin{aligned} &\min_{a,b,c} \sum_{x \in S} f(x) - q(x; a, b, c) + \delta e^T a \\ &\text{s.t.} \quad f(x) \geq q(x; a, b, c), \qquad\qquad \text{for all } x \in S \\ &\qquad\quad a \geq 0. \end{aligned}$$

Since q depends linearly on a, b, c, problem (CGU) is a linear program. The term $\delta e^T a$ reduces the absolute value of the smallest eigenvalue of $\mathrm{Diag}(a)$ in the case where (CGU) is degenerated. The quality of these underestimators depends strongly on the sample set S. If S contains all minimizers of f over $[\underline{x}, \overline{x}]$, the bound is rigorous. Since we cannot guarantee to find all minimizers, this approach provides a heuristic underestimator.

4 Relaxation-Based Heuristic

For computing solution candidates of a MINLP a rounding heuristic was developed, which is based on a convex relaxation of the given problem. The heuristic works by computing a solution point of problem (C), and rounding some binary components of this point. The rounded variables are fixed and the restricted convex relaxation is solved. This procedure is repeated as long, as either the restricted convex relaxation is infeasible, or all binary variables are fixed. In the latter case a local search is started from the last solution of the restricted convex relaxation giving a solution candidate. The values of the binary variables are recursively switched and the whole process is repeated as long, as either all combinations of binary variables are searched, or the number of solution candidates exceeds a given number s_{max}.

To describe this method more detailed, let us define the original problem and the relaxed problem with fixed binary variables

$$(P_{y.J}) \qquad\qquad \min\{c^T x \mid x \in \Omega \cap U_{y,J}\}$$

and

$$(C_{y,J}) \qquad\qquad \min\{c^T x \mid x \in \check{\Omega} \cap U_{y,J}\}$$

where

$$U_{y,J} = \{x \in \mathbb{R}^n \mid x_J = y_J\},$$

$y \in [\underline{x}, \overline{x}]$, $J \subset \{1, \ldots, n\}$, and Ω and $\check{\Omega}$ are the feasible sets of problems (P_{ext}) and (C) respectively. The recursive algorithm described in Fig. 2 computes a set Z of binary feasible solution candidates of problem (P_{ext}).

In order to improve the simple method `Optimize` for solving the continuous nonlinear program ($P_{y,B}$), a neighbourhood search or a deformation heuristic, as described in [AN02], can be used.

5 Preliminary Numerical Results

We tested the described algorithm using a set of instances of MINLP that were obtained from the GAMS web-site [BDM03, CG02]. For the computation of the minimum eigenvalue and corresponding eigenvector we used the Lanczos method ARPACK++ [GS97].

```
RoundHeu:
```

$Z = \emptyset$;
FixVariables$(0, \emptyset)$;

FixVariables(y, J):

if $\check{\Omega} \cap U_{y,J} = \emptyset$, then return;
if $J = B$,
then Optimize(y);
else begin
    get a solution x of $(C_{y,B})$;
    get $K \subset B \setminus J$ where $\min\{x_j - \underline{x}_j, \overline{x}_j - x_j\}$ is small for $j \in K$;
    for $j \in K$, round x_j;
    repeat
            call FixVariables$(x, J \cup K)$;
            switch some components of x_K to get a new binary combination of x_K;
    until ($|Z| \geq s_{\max}$ or all combinations of x_K are searched);
    end

Optimize(y):

if $|B| = n$,
then if $y \in \Omega$, then add y to Z;
else get a solution candidate of $(P_{y,B})$ by using a local search starting from the
    solution of $(C_{y,B})$, and add the point to Z;

Fig. 2. Algorithm *rounding heuristic*

The sequential quadratic programming code SNOPT [GMS97] is used for finding local solutions of nonlinear optimization problems. To analyse its performance, the code was run on a machine with a 1GHz Pentium III processor and 256 MB RAM. Table 1 reports the performance on a selected set of problems using Algorithm RoundHeu. The definition of each column is explained in Table 2. The limit on $|S_f|$, the maximum number of feasible points explored was set to 50.

6 Conclusions

We described the C++ library LaGO for solving general nonconvex MINLPs. Main features of LaGO are: (i) the objective and constraint functions can be given in a black-box formulation, i.e. it is only assumed that functions, gradients and Hessians can be evaluated; (ii) the given MINLP is automatically reformulated as a block-separable formulation giving the possibility to apply decomposition techniques; (iii) the object oriented design of the software allows easy extensions and modifications of the current code.

Table 1. Algorithm performance on a selected set of described examples

example	rel. error	heu. time	k	$\lvert S_f \rvert$	n	m	$\lvert E \rvert$	$\lvert C \rvert$	$\lvert B \rvert$	$\lvert U \rvert$
alan	0	0.02	6	6	9	8	3	5	4	-
batch	0	57.13	84	50	47	74	13	23	24	-
batchdes	0	0.70	12	8	20	20	7	11	9	-
ex1221	0	0.04	4	4	6	6	3	3	3	-
ex1222	0	0.00	1	1	4	4	1	3	1	-
ex1223b	0	0.11	16	16	8	10	1	4	4	-
ex1224	0	13.32	81	45	12	8	3	4	8	-
ex1225	0	0.07	7	6	9	11	3	3	6	-
ex1226	0	0.03	6	5	6	6	2	3	3	-
ex1252	.03	365.73	121	50	40	44	23	25	15	-
gbd	0	0.01	3	3	5	5	1	2	3	-
sep1	0	0.11	3	3	30	32	23	28	2	-
synthes1	0	0.10	5	5	7	7	1	4	3	-
synthes3	0	2.04	20	20	18	24	3	10	8	-
elf	.40	52.22	118	50	55	39	7	31	24	30
ex3	0	1.53	4	4	33	32	18	25	8	15
ex4	0	31.69	22	20	37	31	1	12	25	7
fuel	0	0.56	4	4	16	16	7	13	3	6
gkocis	0	0.08	6	6	12	9	6	9	3	6
meanvarx	0	0.51	20	20	36	45	9	22	14	21
oaer	0	0.05	6	6	10	8	4	7	3	6
procsel	0	0.04	2	2	11	8	5	8	3	6
synheat	0	15.94	21	20	57	65	21	45	12	32
synthes2	0	0.26	12	12	12	15	2	7	5	2
24 examples,	22	solved								

Preliminary results with a relaxation-based rounding heuristic were presented, demonstrating that the current version of LaGO is able to solve medium size MINLP instances in a reasonable time. The only two instances that were not solved to known global optimality were those were the limit of $\lvert S_f \rvert \leq 50$ feasible solutions was reached.

The performance of the described solution approach depends mainly on the quality of the relaxation. In order to improve the relaxation, and to search systematically for a global solution, a branch-and-cut algorithm with box-reduction is currently under development and will be added in the future.

Table 2. Descriptions of the columns of Table 1

rel. error	$|f_{\text{heu}} - f_{\text{best}}|/(1 + |f_{\text{best}}|)$, where f_{heu}, and f_{best} are the objective value obtained with the heuristic and the best known objective value for the problem respectively.
heu. time	seconds required by Algorithm RoundHeu (without preprocessing) to obtain the heuristic solution x_{heu}, and its corresponding objective value f_{heu}.
k	$k \leq 2^{|B|}$ number of fixed binary combinations explored
$|S_f|$	$|S_f| \leq k$, number of feasible solutions obtained from each $(C_{y,B})$
	description of the MINLP instances.
n	number of variables,
m	number of constraints,
$|E|$	number of equality constraints,
$|C|$	number of continuous variables,
$|B|$	number of binary variables,
$|U|$	number of unbounded variables $(U = \{i \mid \underline{x}_i = -\infty \text{ or } \overline{x}_i = +\infty\})$.

References

[ADFN98] C. S. Adjiman, S. Dallwig, C. A. Floudas, and A. Neumaier. A global optimization method, αBB, for general twice-differentiable constrained NLPs — I. Theoretical advances. *Computers and Chemical Engineering*, pages (1137–1158), 1998.

[AF97] C. S. Adjiman and C. A. Floudas. Rigorous convex underestimators for general twice-differentiable problems. *J. of Global Opt.*, 9:23–40, 1997.

[AN02] H. Alperin and I. Nowak. Lagrangian Smoothing Heuristics for MaxCut. Technical report, HU–Berlin NR–2002-6, 2002.

[BDM03] M. R. Bussieck, A. S. Drud, and A. Meeraus. MINLPLib - A Collection of Test Models for Mixed-Iinteger Nonlinear Programming. *INFORMS J. Comput.*, 15(1), 2003.

[BP03] M. R. Bussieck and A. Pruessner. Mixed-integer nonlinear programming. *SIAG/OPT Newsletter: Views & News.*, 2003.

[BSV+01] C. Bliek, P. Spellucci, L. Vicente, A. Neumaier, L. Granvilliers, E. Monfroy, F. Benhamou, E. Huens, P. Van Hentenryck, D. Sam-Haroud, and B. Faltings. *COCONUT Deliverable D1, Algorithms for Solving Nonlinear Constrained and Optimization Problems: The State of The Art.* http://www.mat.univie.ac.at/ neum/glopt.html, 2001.

[CG02] GAMS Development Corp. and GAMS Software GmbH. MINLP World, 2002. http://www.gamsworld.org/minlp/.

[FGK93] Robert Fourer, David M. Gay, and Brian W. Kernighan. *AMPL: A Modeling Language for Mathematical Programming.* Duxbury Press, Brooks/Cole Publishing Company, 1993.

[Flo00] C. A. Floudas. *Deterministic Global Optimization: Theory, Algorithms and Applications.* Kluwer Academic Publishers, 2000.

[GMS97]   P. E. Gill, W. Murray, and M. A. Saunders. SNOPT 5.3 user's guide. Technical report, University of California, San Diego, Mathematics Department Report NA 97-4, 1997.

[GS97]   F. Gomes and D. Sorensen. ARPACK++: a C++ Implementation of ARPACK eigenvalue package, 1997.
http://www.crpc.rice.edu/software/ARPACK/.

[HP95]   R. Horst and P. Pardalos. Handbook of Global Optimization. Kluwer Academic Publishers, 1995.

[KAGB01]   P. Kesavan, R. J. Allgor, E. P. Gatzke, and P. I. Barton. Outer Approximation Algorithms for Separable Nonconvex Mixed-Integer Nonlinear Programs. submitted to Mathematical Programming, 2001.

[Now02]   I. Nowak. Lagrangian Decomposition of Mixed-Integer All-Quadratic Programs. Technical report, HU–Berlin NR–2002–7, 2002.

[PRW95]   A. Phillips, J. Rosen, and V. Walke. Molecular structure determination by global optimization. In Dimacs Series in Discrete Mathematics and Theoretical Computer Science, volume 23, pages 181–198. 1995.

[SP99]   E. M. B. Smith and C. C. Pantelides. A Symbolic Reformulation/Spatial Branch and Bound Algorithm for the Global Optimization of nonconvex MINLPs. Computers and Chemical Engineering, 23:457–478, 1999.

[TS02]   M. Tawarmalani and N. V. Sahinidis. Convexification and Global Optimization in Continuous and Mixed-Integer Nonlinear Programming: Theory, Algorithms, Software, and Applications. Kluwer Academic Publishers, 2002.

[ULP+02]   Zsolt Ugray, Leon Lasdon, John Plummer, Fred Glover, Jim Kelly, and Rafael Marti. A multistart scatter search heuristic for smooth NLP and MINLP problems. http://www.utexas.edu/courses/lasdon/papers.htm, 2002.

[VEH96]   R. Vaidyanathan and M. EL-Halwagi. Global optimization of nonconvex MINLP's by interval analysis. In I. E. Grossmann, editor, Global Optimization in Engineering Design, pages 175–193. Kluwer Academic Publishers, 1996.

[WSV00]   H. Wolkowicz, R. Saigal, and L. Vandenberghe. Handbook of Semidefinite Programming. Kluwer Academic Publishers, 2000.

Solving Global Optimization Problems over Polynomials with GloptiPoly 2.1

Didier Henrion[1,2] and Jean-Bernard Lasserre[1]

[1] Laboratoire d'Analyse et d'Architecture des Systèmes
Centre National de la Recherche Scientifique
7 Avenue du Colonel Roche, 31 077 Toulouse, France
{henrion,lasserre}@laas.fr
[2] Institute of Information Theory and Automation
Academy of Sciences of the Czech Republic
Pod vodárenskou věží 4, 182 08 Praha, Czech Republic
henrion@utia.cas.cz

Abstract. GloptiPoly is a Matlab/SeDuMi add-on to build and solve convex linear matrix inequality relaxations of the (generally non-convex) global optimization problem of minimizing a multivariable polynomial function subject to polynomial inequality, equality or integer constraints. It generates a series of lower bounds monotonically converging to the global optimum. Global optimality is detected and isolated optimal solutions are extracted automatically. In this paper we first briefly describe the theoretical background underlying the relaxations. Following a small illustrative example of the use of GloptiPoly, we then evaluate its performance on benchmark test examples from global optimization, combinatorial optimization and polynomial systems of equations.

1 Introduction

GloptiPoly is a Matlab[1] freeware that builds and solves convex linear matrix inequality (LMI, see [VB96]) relaxations of (generally non-convex) global optimization problems with multivariable real-valued polynomial objective function and constraints. The software solves a series of convex relaxations of increasing size, whose optima are guaranteed to converge monotonically to the global optimum of the original non-convex optimization problem.

GloptiPoly solves LMI relaxations with the help of the semidefinite programming (SDP) solver SeDuMi [SDM99], taking full advantage of sparsity and special problem structure. Optionally, a user-friendly interface called DefiPoly, based on Matlab Symbolic Math Toolbox, can be used jointly with GloptiPoly to define the optimization problems symbolically with a Maple-like syntax.

Our motivations for presenting GloptiPoly to the global optimization community are as follows:

---

[1] Matlab is a trademark of The MathWorks, Inc.

C. Bliek et al. (Eds.): COCOS 2002, LNCS 2861, pp. 43–58, 2003.
© Springer-Verlag Berlin Heidelberg 2003

- We believe that the general methodology behind GloptiPoly (theory of moments, (real) algebraic geometry, LMI relaxations, see section 2) is sufficiently general to find several applications besides those described in this paper. For instance, LMI relaxations can be used to obtain better and better lower bounds in branch and bound procedures for global optimization of polynomial functions;
- As shown in section 3, GloptiPoly is a user-friendly package that researchers and students can experiment easily. It can also be used as a tutorial material to illustrate the use of convex relaxations in global optimization.

GloptiPoly is aimed at small- and medium-scale problems. Numerical experiments in section 5 illustrate that for most of the problem instances available in the literature, the global optimum is reached exactly with LMI relaxations of medium size, at a relatively low computational cost.

GloptiPoly requires Matlab version 5.3 or higher [Mat01], together with the freeware solver SeDuMi version 1.05 [SDM99]. For installation instructions and a comprehensive user's guide, see

$$\texttt{www.laas.fr/}{\sim}\texttt{henrion/software/gloptipoly}$$

2 Theoretical Background

GloptiPoly is based on the theory of positive polynomials and moments described in [Las01, Las02] and briefly summarized in the sequel.

2.1 Introduction

Consider the general nonlinear optimization problem

$$\mathbb{P} \to p^* := \min_{x \in \mathbb{R}^n} \{g_0(x) \mid g_k(x) \geq 0, \ k = 1, \dots m\} \tag{1}$$

where all the $g_k(x) : \mathbb{R}^n \to \mathbb{R}$ are real-valued polynomials of $\mathbb{R}[x_1, \dots, x_n]$. Equality constraints are allowed via two opposite inequalities, so that (1) describes all optimization problems that involve polynomials. In particular, it encompasses non-convex quadratic problems as well as discrete optimization problems (e.g. 0-1 nonlinear programming problems).

The idea behind the methodology of GloptiPoly is to build up a sequence of convex semidefinite relaxations of $\mathbb{P}$ of increasing size and whose sequence of optimal values converges to the global optimal value $p^* = \inf \mathbb{P}$.

The original idea can be traced back to the pioneering Reformulation Linearization Technique (RLT) of [SA90, SA99] where additional redundant constraints (products of the original ones) are introduced and linearized in a higher space (lifting) by introducing additional variables (e.g. $x_i x_j = y_{ij}$) so as to obtain a LP-relaxation. Convergence was proved for 0-1 nonlinear programs. Later, Shor [Sho87, Sho98] also proposed a lifting procedure to reduce any polynomial programming problem to a quadratic one and then use a semidefinite relaxation to obtain a lower bound of p^*, see also the more recent work [Nes00]. Then, the striking certified good approximation of Goemans and Williamson for the MAX-CUT problem [GW95], obtained from a simple SDP (or LMI) relaxation definitely excited the curiosity of researchers for SDP relaxations. However, excepted for the LP-relaxations of Sherali and Adams in 0-1 problems, no proof of convergence was provided.

The proof of convergence of the LMI relaxations defined in [Las01, Las02] and used in GloptiPoly is based on recent results of real algebraic geometry concerning the representation of polynomials, strictly positive on a semi-algebraic set; see also [Par00] for a related approach. It turns out that the primal and dual LMI relaxations of GloptiPoly match both sides of the dual theories of *moments* and *positive polynomials*.

Indeed, while the primal relaxations aim at founding the moments of a probability measure with mass concentrated on some global minimizers of $\mathbb{P}$, the dual relaxations aim at representing the polynomial $g_0(x) - p^*$, nonnegative on the (semi-algebraic) feasible set $\mathbb{K}$ of $\mathbb{P}$, as a linear combination of the g_i's with weights being polynomials that are sums of squares, as in Putinar's representation of polynomials, strictly positive on a semi-algebraic set [Put93].

In brief, the primal LMI relaxations $\{\mathbb{Q}_i\}$ of $\mathbb{P}$ are relaxations of the problem (equivalent to $\mathbb{P}$)

$$p^* = \min_\mu \{ \int g_0 \, d\mu \mid \mu(\mathbb{K}) = 1 \},$$

where the minimum is taken over all the probability measures on the feasible set $\mathbb{K}$ of $\mathbb{P}$, whereas the dual relaxations $\{\mathbb{Q}_i^*\}$ solve

$$\max_{\rho_i, \{q_k\}} \{ \rho_i \mid g_0(x) - \rho_i = q_0 + \sum_{k=1}^m g_k(x) q_k(x) \}, \qquad (2)$$

where the unknowns $\{q_k\}$ are polynomials, all sums of squares, and with degree at most $2i$. For a brief account of these two dual points of view, the interested reader is referred to [Las01, Las02] and the references therein.

The increasing size of the relaxations reflects the effort in the degree $2i$ needed in (2) for ρ_i to be as closed as desired to p^* (and often to be exactly equal to p^*).

2.2   Brief Description of the Methodology

Notation and Definitions. Given any two real-valued symmetric matrices A, B let $\langle A, B \rangle$ denote the usual scalar product trace(AB) and let $A \succeq B$ (resp. $A \succ B$) stand for $A - B$ positive semidefinite (resp. $A - B$ positive definite). Let

$$1, x_1, x_2, \ldots x_n, x_1^2, x_1 x_2, \ldots, x_1 x_n, x_2^2, x_2 x_3, \ldots, x_n^2, \ldots, x_1^r, \ldots, x_n^r, \qquad (3)$$

be a basis for the space $\mathcal{A}_r$ of real-valued polynomials of degree at most r, and let $s(r)$ be its dimension. Therefore, a polynomial $p : \mathbb{R}^n \to \mathbb{R}$ of degree r is written

$$p(x) = \sum_\alpha p_\alpha x^\alpha, \qquad x \in \mathbb{R}^n,$$

where

$$x^\alpha = x_1^{\alpha_1} x_2^{\alpha_2} \dots x_n^{\alpha_n}, \quad \text{with} \quad \sum_{i=1}^n \alpha_i = k,$$

is a monomial of degree k with coefficient p_α. Let $p = \{p_\alpha\} \in \mathbb{R}^{s(r)}$ be the vector of coefficients of the polynomial $p(x)$ in the basis (3).

Given an $s(2r)$-sequence $(1, y_1, \dots,)$, let $M_r(y)$ be the moment matrix of dimension $s(r)$ with rows and columns indexed by (3). For instance, to fix ideas, consider the 2-dimensional case. The moment matrix $M_r(y)$ is the block matrix $\{M_{i,j}(y)\}_{0 \le i,j \le r}$ defined by

$$M_{i,j}(y) = \begin{bmatrix} y_{i+j,0} & y_{i+j-1,1} & \cdots & y_{i,j} \\ y_{i+j-1,1} & y_{i+j-2,2} & \cdots & y_{i-1,j+1} \\ \cdots & \cdots & \cdots & \cdots \\ y_{j,i} & y_{i+j-1,1} & \cdots & y_{0,i+j} \end{bmatrix}.$$

Thus, with $n = 2$ and $r = 2$, one obtains

$$M_2(y) = \begin{bmatrix} 1 & y_{10} & y_{01} & y_{20} & y_{11} & y_{0,2} \\ y_{10} & y_{20} & y_{11} & y_{30} & y_{21} & y_{12} \\ y_{01} & y_{11} & y_{02} & y_{21} & y_{12} & y_{03} \\ y_{20} & y_{30} & y_{21} & y_{40} & y_{31} & y_{22} \\ y_{11} & y_{21} & y_{12} & y_{31} & y_{22} & y_{13} \\ y_{02} & y_{12} & y_{03} & y_{22} & y_{13} & y_{04} \end{bmatrix}$$

Another, more intuitive way of constructing $M_r(y)$ is as follows. If $M_r(y)(1, i) = y_\alpha$ and $M_r(y)(j, 1) = y_\beta$, then $M_r(y)(i, j) = y_{\alpha+\beta}$, with $\alpha + \beta = (\alpha_1 + \beta_1, \cdots, \alpha_n + \beta_n)$. This defines a bilinear form $\langle ., . \rangle_y$ on $\mathcal{A}_r$, by $\langle q(x), v(x) \rangle_y := \langle q, M_r(y)v \rangle$, $q(x), v(x) \in \mathcal{A}_r$, and if y is a sequence of moments of some measure μ_y, then

$$\langle q, M_r(y)q \rangle = \int q(x)^2 \, \mu_y(dx) \ge 0, \tag{4}$$

so that $M_r(y) \succeq 0$.

If the entry (i, j) of the matrix $M_r(y)$ is y_β, let $\beta(i, j)$ denote the subscript β of y_β. Next, given a polynomial $\theta(x) : \mathbb{R}^n \to \mathbb{R}$ with coefficient vector θ, we define the matrix $M_r(\theta y)$ by

$$M_r(\theta y)(i, j) = \sum_\alpha \theta_\alpha y_{\{\beta(i,j)+\alpha\}}. \tag{5}$$

For instance, with

$$M_1(y) = \begin{bmatrix} 1 & y_{10} & y_{01} \\ y_{10} & y_{20} & y_{11} \\ y_{01} & y_{11} & y_{02} \end{bmatrix} \quad \text{and} \quad x \mapsto \theta(x) := a - x_1^2 - x_2^2,$$

we obtain

$$M_1(\theta y) = \begin{bmatrix} a - y_{20} - y_{02} & ay_{10} - y_{30} - y_{12} & ay_{01} - y_{21} - y_{03} \\ ay_{10} - y_{30} - y_{12} & ay_{20} - y_{40} - y_{22} & ay_{11} - y_{31} - y_{13} \\ ay_{01} - y_{21} - y_{03} & ay_{11} - y_{31} - y_{13} & ay_{02} - y_{22} - y_{04} \end{bmatrix}.$$

In a manner similar to what we have in (4), if y is a sequence of moments of some measure μ_y, then

$$\langle q, M_r(\theta y)q \rangle = \int \theta(x)q(x)^2 \, \mu_y(dx),$$

for every polynomial $q(x) : \mathbb{R}^n \to \mathbb{R}$ with coefficient vector $q \in \mathbb{R}^{s(r)}$. Therefore, $M_r(\theta y) \succeq 0$ whenever μ_y has its support contained in the set $\{\theta(x) \geq 0\}$. The matrix $M_r(\theta y)$ is called a *localizing* matrix.

The $\mathbb{K}$-moment problem identifies those sequences y that are moment-sequences of a measure with support contained in the semi-algebraic set $\mathbb{K}$. In duality with the theory of moments is the theory of representation of positive polynomials, which dates back to Hilbert's 17th problem. This fact will be reflected in the semidefinite relaxations proposed later.

LMI Relaxations. Let $\mathbb{P}$ be the problem defined in (1) and let

$$\mathbb{K} := \{x \in \mathbb{R}^n \,|\, g_k(x) \geq 0, \; k = 1, \ldots, m\} \tag{6}$$

be the feasible set associated with $\mathbb{P}$.

Depending on its parity, let $\text{degree}(g_k) = 2v_k - 1$ or $2v_k$, for all $k = 0, 1, \ldots, m$. When needed below, for $i \geq \max_k v_k$, the vectors $g_k \in \mathbb{R}^{s(2v_k)}$ are extended to vectors of $\mathbb{R}^{s(2i)}$ by completing with zeros. As we minimize $g_0(x)$ we may and will assume that its constant term is zero, that is, $g_0(0) = 0$.

For $i \geq \max_{k \in \{0,m\}} v_k$, consider the following family $\{\mathbb{Q}_i\}$ of convex positive semidefinite programs, or LMI relaxations of $\mathbb{P}$:

$$\mathbb{Q}_i \begin{cases} \displaystyle\min_y \; \sum_\alpha (g_0)_\alpha y_\alpha \\ \qquad M_i(y) \succeq 0 \\ \qquad M_{i-v_k}(g_k y) = 0, \quad k = 1, \ldots, m, \end{cases}$$

with respective dual problems

$$\mathbb{Q}_i^* \begin{cases} \displaystyle\min_{X \succeq 0, Z_k} \; -X(1,1) - \sum_{k=1}^m g_k(0)Z_k(1,1) \\ \qquad \langle X, B_\alpha \rangle + \sum_{k=1}^m \langle Z_k, C_\alpha^k \rangle = (g_0)_\alpha, \; \forall \alpha \neq 0 \end{cases}$$

where X, Z_k are real-valued symmetric matrices, the "dual variables" associated with the constraints $M_i(y) \succeq 0$ and $M_{i-v_k}(g_k y) \succeq 0$ respectively, and where we have written

$$M_i(y) = \sum_\alpha B_\alpha y_\alpha; \; M_{i-v_k}(g_k y) = \sum_\alpha C_\alpha^k y_\alpha, \; k = 1, \ldots, n,$$

for appropriate real-valued symmetric matrices $B_\alpha, C_\alpha^k, \; k = 1, \ldots, n$.

In the standard terminology, the constraint $M_i(y) \succeq 0$ is called a linear matrix inequality (LMI) and $\mathbb{Q}_i$ and its dual $\mathbb{Q}_i^*$ are so-called positive semidefinite programs, the LMI relaxations of $\mathbb{P}$. The reader interested in more details on SDP and LMIs is referred to [VB96] and the many references therein.

Remark. In the case of 0-1 programming (and more generally, discrete optimization problems) the relaxations $\mathbb{Q}_i$ simplify. Indeed, instead of explicitly stating the LMIs associated with the integrality constraints $x_i^2 = 1$, $i = 1, ..., n$, it suffices to replace (in all the other LMIs) every occurrence of a variable y_α by y_β with $\beta_i = 1$ if $\alpha_i >= 1$, for all $i = 1, ..., n$. This significantly reduces the number of variables in the resulting relaxation $\mathbb{Q}_i$.

Convergence. We make the following assumption on the set $\mathbb{K}$ defined in (6).

Assumption A. $\mathbb{K}$ is compact and there exist a polynomial $u \in \mathbb{R}[x_1, \ldots, x_n]$ such that

- the set $\{x \in \mathbb{R}^n \mid u(x) \geq 0\}$ is compact;
- polynomial $u(x)$ can be written as

$$q_0(x) + \sum_{i=1}^{m} q_i(x)g_i(x) \qquad (7)$$

for some polynomials $\{q_i\}$, all sums of squares.

Assumption A is satisfied in many cases of interest. For instance, it holds as soon as $\{x \in \mathbb{R}^n \mid g_k(x) \geq 0\}$ is compact for some $k \in \{1, \ldots, m\}$, or when all the g_i are linear and $\mathbb{K}$ is compact (hence a convex polytope), for 0-1 (and more generally discrete) programs. In addition, if one knows that a global minimizer x^* of $\mathbb{P}$ satisfies $\|x^*\| \leq M$ for some $M > 0$, then adding the constraint $M - \|x\| \geq 0$ in the definition (6) of $\mathbb{K}$ will ensure that Assumption A holds.

Under Assumption A it was proved in [Las01] that $\inf \mathbb{Q}_i \uparrow \inf \mathbb{P}$ as $i \to \infty$. Moreover, if $g_0(x) - p^*$ has the representation (7) for some polynomials $\{q_i\}$, all sums of squares, and of degree at most $2i_0$, then for all $j \geq i_0$,

$$\sup \mathbb{Q}_j^* = \max \mathbb{Q}_j^* = \min \mathbb{Q}_j = \inf \mathbb{Q}_j = \min \mathbb{P} = p^*.$$

In addition, any optimal solution of $\mathbb{Q}_j$ identifies the vector of moments of a probability measure with mass concentrated on some global minimizers of $\mathbb{P}$.

3   Illustration

In this section we describe a small numerical examples to illustrate the basic use of GloptiPoly. We consider non-convex quadratic problem [Flo99, Pb. 3.5]:

$$\min -2x_1 + x_2 - x_3$$
$$\text{s.t. } x_1(4x_1 - 4x_2 + 4x_3 - 20) + x_2(2x_2 - 2x_3 + 9) + x_3(2x_3 - 13) + 24 \geq 0$$
$$x_1 + x_2 + x_3 \leq 4, \quad 3x_2 + x_3 \leq 6$$
$$0 \leq x_1 \leq 2, \quad 0 \leq x_2, \quad 0 \leq x_3 \leq 3.$$

To define this problem with GloptiPoly we use the following Matlab script:

```
>> P = defipoly({'min -2*x1+x2-x3',...
'x1*(4*x1-4*x2+4*x3-20)+x2*(2*x2-2*x3+9)+x3*(2*x3-13)+24>=0',...
'x1+x2+x3<=4', '3*x2+x3<=6',...
'0<=x1', 'x1<=2', '0<=x2', '0<=x3', 'x3<=3'}, 'x1,x2,x3');
```

In the above script, we use features of the Symbolic Math Toolbox version 2.1, the Matlab gateway to the kernel of Maple V [Map01]. It is also possible to enter problems into GloptiPoly without symbolic computations, see [Glo02] for more information.

To solve the first LMI relaxation of the quadratic problem, we type:

```
>> output = gloptipoly(P)
output =
     status: 0
       crit: -6.0000
        sol: {}
```

Field `status` = 0 indicates that it is not possible to detect global optimality with this LMI relaxation, hence `crit` = -6.0000 is a lower bound on the global optimum.

Next we try to solve the second, third and fourth LMI relaxations of the quadratic problem with the instructions:

```
>> output = gloptipoly(P,2)          >> output = gloptipoly(P,3)
output =                             output =
     status: 0                            status: 0
       crit: -5.6923                        crit: -4.0685
        sol: {}                              sol: {}
>> output = gloptipoly(P,4)
output =
     status: 1
       crit: -4.0000
        sol: {[3x1 double]   [3x1 double]}
>> output.sol{:}
ans =                                ans =
     2.0000                               0.5000
     0.0000                               0.0000
     0.0000                               3.0000
```

Both the second and third LMI relaxations return tighter lower bounds on the global optimum. Eventually global optimality is reached at the fourth LMI relaxation (certified by `status` = 1). GloptiPoly also returns two globally optimal solutions $x_1 = 2$, $x_2 = 0$, $x_3 = 0$ and $x_1 = 0.5$, $x_2 = 0$, $x_3 = 3$ leading to `crit` = -4.0000.

As shown below, the number of LMI variables and the size of the relaxed LMI problem, hence the overall computational time, increase quickly with the relaxation order:

LMI order	1	2	3	4	5	6
LMI optimum	-6.0000	-5.6923	-4.0685	-4.0000	-4.0000	-4.0000
LMI variables	9	34	83	164	285	454
LMI size	24	228	1200	4425	12936	32144

4   Features

As shown by the above numerical example, GloptiPoly is designed to solve an LMI relaxation of a given order, so it can be invoked iteratively with increasing orders until the global optimum is reached. Asymptotic convergence of the optimal values of the LMI relaxations to the global optimal value of the original problem is ensured when the compact set $\mathbb{K}$ of feasible solutions satisfies Assumption A. This condition is satisfied in many practical optimization problems, see [Las01, Las02].

General features of GloptiPoly are listed below:

− Certificate of global optimality
− Automatic extraction of globally optimal solutions
− 0-1 or ±1 integer constraints on some of the decision variables (combinatorial optimization problems)
− Generation of input and output data in SeDuMi's format
− Generation of moment matrices associated with LMI relaxations
− User-defined scaling of decision variables
− Exploits sparsity of polynomial data.

Finally note that for technical reasons there is currently a limitation on the number of variables handled by GloptiPoly. For example, the current version of GloptiPoly is not able to handle quadratic problems with more than 19 variables. This limitation should be removed soon. For more details, see [Glo02].

5   Performance

All the computations in this section were carried out with Matlab 6.1 and SeDuMi 1.05 with relative accuracy `pars.eps = 1e-9` on a PC with a Pentium IV 1.6 Mhz processor with 512 Mb RAM.

A salient feature of these numerical tests is that whenever the global optimum is reached at some LMI relaxation (which is the case in most examples), then the order of this LMI relaxation is relatively low.

5.1   Continuous Optimization Problems

We report in Table 1 the performance of GloptiPoly on a series of benchmark non-convex continuous optimization problems. For each problem we indicated the number of decision variables 'var', the number of inequality or equality constraints 'cstr', and the maximum degree arising in the polynomial expressions 'deg'. In almost all reported instances the global optimum was reached exactly by an LMI relaxation of small order, reported in the column entitled 'order'. CPU times are in seconds. 'LMI var' is the dimension of SeDuMi dual vector y, whereas 'LMI size' is the dimension of SeDuMi primal vector x, see [SDM99]. As indicated by the label 'dim' in the rightmost column, quadratic problems 2.8, 2.9 and 2.11 in [Flo99] involve more than 19 variables and could not be handled by the current version of GloptiPoly. Except for problems 2.4 and 3.2, the computational load is moderate.

5.2   Discrete Optimization Problems

We also report the performance of GloptiPoly on a series of small-size combinatorial optimization problems. In Table 2 we first let GloptiPoly converge to the global optimum, in general extracting several solutions. The number of extracted solutions is reported in the column entitled 'sol'.

Then, we slightly perturbed the criterion to be optimized in order to destroy the problem symmetry. Proceeding this way, the optimum solution is generically unique and convergence to the global optimum is ensured more easily, cf. Table 3. See also [Glo02] for more details on this technique.

5.3   Polynomial Systems of Equations

Multivariate polynomial systems of equations can be solved with GloptiPoly. We tested its performance on a series of benchmark examples taken from [Ver99] and [Fri00], where we removed examples featuring complex coefficients (recall that GloptiPoly handles real-valued polynomials only). Short descriptions of the benchmarks are given in Tables 4 and 5.

We carried out our experiments by solving feasibility problems, i.e. no criterion was optimized. We did not attempt to count or enumerate all the solutions to the polynomial systems of equations, since this is outside the scope of GloptiPoly. Note that in the absence of a criterion to optimize, GloptiPoly solves the LMI relaxations by minimizing the trace of the moment matrix. Alternative criteria (such as e.g. minimum coordinate or minimum Euclidean-norm solution) are of course possible, but not investigated here.

Our results are reported in Tables 6 and 7. Column 'sol' indicates the number of solutions successfully extracted by GloptiPoly. In the last column the label 'mem' means that the error message 'out of memory' was issued by SeDuMi. GloptiPoly successfully solved about 90% of the systems.

6   Conclusion

GloptiPoly is as a general-purpose software with a user-friendly interface to solve in a unified way a wide range of small- to medium-size non-convex polynomial optimization problems. As illustrated by extensive numerical examples, the main strength of GloptiPoly is that no expert tuning is necessary to cope with very distinct problems coming from different branches of engineering and applied mathematics. GloptiPoly can be used as a black-box software, so it cannot be considered as a competitor to highly specialized codes for solving e.g. sparse polynomial systems of equations or large combinatorial optimization problems.

It is well-known that problems involving polynomial bases with monomials of increasing powers are naturally badly conditioned. If lower and upper bounds on the optimization variables are available as problem data, it may be a good idea to scale all the intervals around one. Alternative bases such as Chebyshev polynomials may also prove useful.

Finally, it would be instructive to compare GloptiPoly with the recently developed software SOSTOOLS [SOS02], also invoking SeDuMi to solve sums of squares optimization programs over polynomials, based on the theory described in [Par00].

Acknowledgment

Work of the first author was partially supported by the Grant Agency of the Czech Republic under Project No. 102/02/0709.

References

[Anj01]   M. Anjos. New Convex Relaxations for the Maximum Cut and VLSI Layout Problems. *PhD Thesis*, Waterloo University, Ontario, Canada, 2001. See `orion.math.uwaterloo.ca/~hwolkowi`.

[Flo99]   C. A. Floudas, P. M. Pardalos, C. S. Adjiman, W. R. Esposito, Z. H. Gümüs, S. T. Harding, J. L. Klepeis, C. A. Meyer, C. A. Schweiger. Handbook of Test Problems in Local and Global Optimization. *Kluwer Academic Publishers*, Dordrecht, 1999. See `titan.princeton.edu/TestProblems`.

[Fri00]   The Numerical Algorithms Group Ltd. FRISCO - A Framefork for Integrated Symbolic/Numeric Computation. European Commission Project No. 21-024, Esprit Reactive LTR Scheme, 2000. See `www.nag.co.uk/projects/frisco.html`.

[GW95]   M. X. Goemans, D. P. Williamson. Improved approximation algorithms for maximum cut and satisfiability problems using semidefinite programming. *Journal of the ACM*, Vol. 42, pp. 1115–1145, 1995.

[Glo02]   D. Henrion, J. B. Lasserre. GloptiPoly: Global Optimization over Polynomials with Matlab and SeDuMi. *ACM Transactions on Mathematical Software*, Vol. 29, No. 2, pp. 165-194, 2003. Available at `www.laas.fr/~henrion/software/gloptipoly`.

[Las01]   J. B. Lasserre. Global Optimization with Polynomials and the Problem of Moments. *SIAM Journal on Optimization*, Vol. 11, No. 3, pp. 796–817, 2001.

[Las02]   J. B. Lasserre. An Explicit Equivalent Positive Semidefinite Program for 0-1 Nonlinear Programs. *SIAM Journal on Optimization*, Vol. 12, No. 3, pp. 756–769, 2002.

[Map01]   Waterloo Maple Software Inc. Maple V release 5. 2001. See `www.maplesoft.com`.

[Mat01]   The MathWorks Inc. Matlab version 6.1. 2001. See `www.mathworks.com`.

[Nes00]   Y. Nesterov. Squared functional systems and optimization problems. Chapter 17, pp. 405–440 in H. Frenk, K. Roos, T. Terlaky (Editors). High performance optimization. *Kluwer Academic Publishers*, Dordrecht, 2000.

[Par00]   P. A. Parrilo. Structured Semidefinite Programs and Semialgebraic Geometry Methods in Robustness and Optimization. *PhD Thesis*, California Institute of Technology, Pasadena, California, 2000. See `www.control.ethz.ch/~parrilo`.

[Put93]   M. Putinar. Positive polynomials on compact semi-algebraic sets. *Indiana University Mathematics Journal*, Vol. 42, pp. 969–984, 1993.

[SA90]   H. D. Sherali, W. P. Adams. A hierarchy of relaxations between the continuous and convex hull representations for zero-one programming problems. *SIAM Journal on Discrete Mathematics*, Vol. 3, pp. 411–430, 1990.

[SA99]   H. D. Sherali, W. P. Adams. A reformulation-linearization technique for solving discrete and continuous nonconvex problems, *Kluwer Academic Publishers*, Dordrecht, 1999.

[Sho87]   N. Z. Shor. Quadratic optimization problems. *Tekhnicheskaya Kibernetika*, Vol. 1, pp. 128–139, 1987.

[Sho98]   N. Z. Shor. Nondifferentiable Optimization and Polynomial Problems. *Kluwer Academic Publishers*, Dordrecht, 1998.

[SDM99]   J. F. Sturm. Using SeDuMi 1.02, a Matlab Toolbox for Optimization over Symmetric Cones. *Optimization Methods and Software*, Vol. 11-12, pp. 625–653, 1999. Version 1.05 available at `fewcal.kub.nl/sturm/software/sedumi.html`.

[SOS02]   S. Prajna, A. Papachristodoulou, P. A. Parrilo. SOSTOOLS: Sum of Squares Optimization Toolbox for Matlab. California Institute of Technology, Pasadena, USA, 2002. Version 1.00 available at `www.cds.caltech.edu/sostools`.

[VB96]   L. Vandenberghe, S. Boyd. Semidefinite programming. *SIAM Review*, Vol. 38, pp. 49–95, 1996.

[Ver99]   J. Verschelde. Algorithm 795: PHCpack: A general-purpose solver for polynomial systems by homotopy continuation. *ACM Transactions on Mathematical Software*, Vol. 25, No. 2, pp. 251–276, 1999. Database of polynomial systems available at `www.math.uic.edu/~jan/demo.html`.

Table 1. Continuous optimization problems. CPU times and LMI relaxation orders required to reach global optima

problem	var	cstr	deg	LMI var	LMI size	CPU	order
[Las01, Ex. 1]	2	0	4	14	36	0.13	2
[Las01, Ex. 2]	2	0	4	14	36	0.13	2
[Las01, Ex. 3]	2	0	6	152	2025	1.13	8
[Las01, Ex. 5]	2	3	2	14	63	0.22	2
[Flo99, Pb. 2.2]	5	11	2	461	7987	11.8	3
[Flo99, Pb. 2.3]	6	13	2	209	1421	1.86	2
[Flo99, Pb. 2.4]	13	35	2	2379	17885	1012	2
[Flo99, Pb. 2.5]	6	15	2	209	1519	1.58	2
[Flo99, Pb. 2.6]	10	31	2	1000	8107	67.7	2
[Flo99, Pb. 2.7]	10	25	2	1000	7381	75.3	2
[Flo99, Pb. 2.8]	20	10	2	-	-	-	dim
[Flo99, Pb. 2.9]	24	10	2	-	-	-	dim
[Flo99, Pb. 2.10]	10	11	2	1000	5632	45.3	2
[Flo99, Pb. 2.11]	20	10	2	-	-	-	dim
[Flo99, Pb. 3.2]	8	22	2	3002	71775	3032	3
[Flo99, Pb. 3.3]	5	16	2	125	1017	1.20	2
[Flo99, Pb. 3.4]	6	16	2	209	1568	1.50	2
[Flo99, Pb. 3.5]	3	8	2	164	4425	2.42	4
[Flo99, Pb. 4.2]	1	2	6	6	34	0.17	3
[Flo99, Pb. 4.3]	1	2	50	50	1926	0.94	25
[Flo99, Pb. 4.4]	1	2	5	6	34	0.25	3
[Flo99, Pb. 4.5]	1	2	4	4	17	0.14	2
[Flo99, Pb. 4.6]	2	2	6	27	172	0.41	3
[Flo99, Pb. 4.7]	1	2	6	6	34	0.20	3
[Flo99, Pb. 4.8]	1	2	4	4	17	0.16	2
[Flo99, Pb. 4.9]	2	5	4	14	73	0.31	2
[Flo99, Pb. 4.10]	2	6	4	44	697	0.58	4

Table 2. Discrete optimization problems. CPU times and LMI relaxation orders required to reach global optima and extract several solutions

problem	var	cstr	deg	LMI var	LMI size	CPU	order	sol
QP [Flo99, Pb. 13.2.1.1]	4	4	2	10	29	0.10	1	1
QP [Flo99, Pb. 13.2.1.2]	10	0	2	385	3136	3.61	2	1
Max-Cut P_1 [Flo99, Pb. 11.3]	10	0	2	847	30976	38.1	3	10
Max-Cut P_2 [Flo99, Pb. 11.3]	10	0	2	847	30976	43.7	3	2
Max-Cut P_3 [Flo99, Pb. 11.3]	10	0	2	847	30976	43.0	3	2
Max-Cut P_4 [Flo99, Pb. 11.3]	10	0	2	847	30976	38.8	3	2
Max-Cut P_5 [Flo99, Pb. 11.3]	10	0	2	-	-	-	4	dim
Max-Cut P_6 [Flo99, Pb. 11.3]	10	0	2	847	30976	43.0	3	2
Max-Cut P_7 [Flo99, Pb. 11.3]	10	0	2	847	30976	44.3	3	4
Max-Cut P_8 [Flo99, Pb. 11.3]	10	0	2	847	30976	43.4	3	2
Max-Cut P_9 [Flo99, Pb. 11.3]	10	0	2	847	30976	49.3	3	6
Max-Cut cycle C_5 [Anj01]	5	0	2	31	676	0.19	3	10
Max-Cut complete K_5 [Anj01]	5	0	2	31	961	0.19	4	20
Max-Cut 5-node [Anj01]	5	0	2	31	676	0.24	3	6
Max-Cut antiweb AW_9^2 [Anj01]	9	0	2	-	-	-	4	dim
Max-Cut 10-node Petersen [Anj01]	10	0	2	847	30976	39.6	3	10
Max-Cut 12-node [Anj01]	12	0	2	-	-	-	3	dim

Table 3. Discrete optimization problems. CPU times and LMI relaxation orders required to reach global optima with perturbed criterion

problem	var	cstr	deg	LMI var	LMI size	CPU	order
QP [Flo99, Pb. 13.2.1.1]	4	4	2	10	29	0.06	1
QP [Flo99, Pb. 13.2.1.2]	10	0	2	847	30976	40.0	3
Max-Cut P_1 [Flo99, Pb. 11.3]	10	0	2	385	3136	3.10	2
Max-Cut P_2 [Flo99, Pb. 11.3]	10	0	2	385	3136	3.03	2
Max-Cut P_3 [Flo99, Pb. 11.3]	10	0	2	385	3136	3.98	2
Max-Cut P_4 [Flo99, Pb. 11.3]	10	0	2	385	3136	3.70	2
Max-Cut P_5 [Flo99, Pb. 11.3]	10	0	2	385	3136	3.41	2
Max-Cut P_6 [Flo99, Pb. 11.3]	10	0	2	385	3136	3.66	2
Max-Cut P_7 [Flo99, Pb. 11.3]	10	0	2	385	3136	3.70	2
Max-Cut P_8 [Flo99, Pb. 11.3]	10	0	2	385	3136	3.33	2
Max-Cut P_9 [Flo99, Pb. 11.3]	10	0	2	385	3136	4.03	2
Max-Cut cycle C_5 [Anj01]	5	0	2	30	256	0.22	2
Max-Cut complete K_5 [Anj01]	5	0	2	31	676	0.28	3
Max-Cut 5-node [Anj01]	5	0	2	30	256	0.22	2
Max-Cut antiweb AW_9^2 [Anj01]	9	0	2	465	16900	12.5	3
Max-Cut 10-node Petersen [Anj01]	10	0	2	385	3136	3.14	2
Max-Cut 12-node [Anj01]	12	0	2	793	6241	29.2	2

Table 4. Short descriptions of polynomial systems of equations. Part 1

problem	short description
boon	neurophysiology problem
bifur	non-linear system bifurcation
brown	Brown's 5-dimensional almost linear system
butcher	Butcher's system from PoSSo test suite
camera1s	displacement of camera between two positions
caprasse	Caprasse's system from PoSSo test suite
cassou	Cassou-Nogues's system from PoSSo test suite
chemequ	chemical equilibrium of hydrocarbon combustion
cohn2	Cohn's modular equations for special algebraic number fields
cohn3	Cohn's modular equations for special algebraic number fields
comb3000	combustion chemistry example for a temperature of 3000 degrees
conform1	Emiris' conformal analysis of cyclic molecules ($b_{11} = -9$)
conform2	Emiris' conformal analysis of cyclic molecules ($b_{11} = -\sqrt{3}/2$)
conform3	Emiris' conformal analysis of cyclic molecules ($b_{11} = -310$)
conform4	Emiris' conformal analysis of cyclic molecules ($b_{11} = -13$)
cpdm5	5-dimensional system of Caprasse and Demaret
d1	sparse system by Hong and Stahl
des18_3	dessin d'enfant
des22_24	dessin d'enfant
discret3	from PoSSo test suite
eco5	5-dimensional economics problem
eco6	6-dimensional economics problem
eco7	7-dimensional economics problem
eco8	8-dimensional economics problem
fourbar	four-bar mechanical design problem
geneig	generalized eigenvalue problem
heart	heart dipole problem
i1	interval arithmetic benchmark
ipp	six-revolute-joint problem of mechanics
katsura5	problem of magnetism in physics
kinema	robot kinematics problem
kin1	inverse kinematics of an elbow manipulator
ku10	10-dimensional system of Ku
lorentz	equilibrium points of 4-dimensional Lorentz attractor
manocha	intersection of high-degree polynomial curves
noon3	neural network modeled by adaptive Lotka-Volterra system
noon4	neural network modeled by adaptive Lotka-Volterra system
noon5	neural network modeled by adaptive Lotka-Volterra system
proddeco	system with product-decomposition structure
puma	hand position and orientation of PUMA robot
quadfor2	Gaussian quadrature formula with 2 knots and 2 weights over [-1,+1]
quadgrid	interpolating quadrature formula for function defined on a grid

Table 5. Short descriptions of polynomial systems of equations. Part 2

problem	short description
rabmo	optimal multi-dimensional quadrature formulas
rbpl	generic positions of parallel robot
redeco5	reduced 5-dimensional economics problem
redeco6	reduced 6-dimensional economics problem
redeco7	reduced 7-dimensional economics problem
redeco8	reduced 8-dimensional economics problem
rediff3	3-dimensional reaction-diffusion problem
reimer5	5-dimensional system of Reimer
rose	general economic equilibrium problem
s9_1	small system from constructive Galois theory
sendra	from PoSSo test suite
solotarev	from PoSSo test suite
stewart1	direct kinematic problem of parallel robot
stewart2	direct kinematic problem of parallel robot
trinks	from PoSSo test suite
virasoro	construction of Virasoro algebras
wood	system derived from optimizing the Wood function
wright	Wright's system

Table 6. Polynomial systems of equations. CPU times and LMI relaxation orders required to reach global optima. Part 1

problem	var	cstr	deg	LMI var	LMI size	CPU	order	sol
boon	6	6	4	3002	52864	1220	4	8
bifur	3	3	9	454	8717	8.20	5	2
brown	5	5	5	461	4061	6.27	3	1
butcher	7	7	4	6434	120156	-	4	mem
camera1s	6	6	2	209	952	1.33	2	2
caprasse	4	4	4	209	1285	0.58	3	2
cassou	4	4	8	4844	280151	-	8	mem
chemequ	5	5	3	461	3661	9.48	3	1
chemequs	5	5	3	124	486	6.73	2	1
cohn2	4	4	6	209	1229	0.48	3	1
cohn3	4	4	6	209	1229	0.55	3	1
comb3000	10	10	3	1000	4951	24.6	2	1
conform1	3	3	4	83	430	0.22	3	2
conform2	3	3	4	83	430	0.19	3	2
conform3	3	3	4	285	3766	3.89	5	4
conform4	3	3	4	454	8946	12.2	6	2
cpdm5	5	5	3	125	446	0.24	2	1
d1	12	12	3	-	-	-	3	dim
des18_3	8	8	3	12869	303945	-	4	mem
des22_24	10	10	2	1000	5016	77.2	1	1
discret3	8	8	2	44	89	0.31	1	1

Table 7. Polynomial systems of equations. CPU times and LMI relaxation orders required to reach global optima. Part 2

problem	var	cstr	deg	LMI vas	LMI size	CPU	order	sol
eco5	5	5	3	461	3661	5.98	3	1
eco6	6	6	3	923	7980	57.4	3	1
eco7	7	7	3	1715	15921	256	3	1
eco8	8	8	3	3002	29565	1310	3	1
fourbar	4	4	4	69	229	0.16	2	1
geneig	6	6	3	923	7602	33.2	3	1
heart	8	8	4	3002	31545	1532	3	2
i1	10	10	3	1000	4366	44.1	2	1
ipp	8	8	2	494	2385	6.42	2	1
katsura5	6	6	2	209	952	0.74	2	1
kinema	9	9	2	714	3520	26.4	2	1
kin1	12	12	3	-	-	-	3	dim
ku10	10	10	2	1000	5016	72.5	2	1
lorentz	4	4	2	209	1705	0.64	2	2
manocha	2	2	8	90	826	1.27	6	1
noon3	3	3	3	83	430	0.22	3	1
noon4	4	4	3	209	1285	0.65	3	1
noon5	5	5	3	461	3241	4.48	3	1
proddeco	4	4	4	69	229	0.11	2	1
puma	8	8	2	3002	35505	1136	3	4
quadfor2	4	4	4	209	1495	0.75	3	2
quadgrid	5	5	5	461	3641	10.52	3	1
rabmo	9	9	5	5004	51703	-	3	mem
rbpl	6	6	3	923	7602	36.9	3	1
redeco5	5	5	2	20	41	0.16	1	1
redeco6	6	6	2	27	55	0.13	1	1
redeco7	7	7	2	35	71	0.14	1	1
redeco8	8	8	2	44	89	0.13	1	1
rediff3	3	3	2	9	19	0.09	1	1
reimer5	5	5	6	6187	264516	-	6	mem
rose	3	3	9	679	16681	79.5	7	2
s9_1	8	8	2	494	2385	5.45	2	1
sendra	2	2	7	65	453	0.34	5	1
solotarev	4	4	3	69	257	0.24	2	1
stewart1	9	9	2	714	3520	20.4	2	2
stewart2	12	10	2	1819	9191	372	2	1
trinks	6	6	3	209	925	0.78	2	1
virasoro	8	8	2	44	89	0.16	1	1
wood	4	3	2	69	527	0.20	2	1
wright	5	5	2	20	41	0.17	1	1

Rigorous Error Bounds for the Optimal Value of Linear Programming Problems

Christian Jansson

Inst. of Computer Science III, Technical University Hamburg-Harburg
Schwarzenbergstraße 95, 21071 Hamburg, Germany
jansson@tu-harburg.de
Tel.: +49 40 428783887

Keywords: linear programming, interval arithmetic, rigorous error bounds, sensitivity analysis, branch-and-bound

1 Introduction

We consider the computation of rigorous lower and upper error bounds for the optimal value in linear programming.

During the last decades linear programming problems have been solved very successfully. Sometimes, due to floating point arithmetic, the computed results may be wrong, especially for ill-conditioned problems. Ill-conditioning is not a rare phenomenon. In a recent paper Ordóñez and Freund [13] have pointed out that many lp-instances in the NETLIB Linear Programming Library [10] are ill-conditioned. Hence, rigorous error bounds may be useful in linear programming.

In global optimization such bounds can be used also. There, frequently linear relaxations are solved sequentially within branch and bound frameworks. In order to discard subproblems containing no global optimal points, safe lower bounds for the optimal value of linear relaxations are required. For comprehensive treatments of relaxation techniques for mixed integer nonlinear programming problems, see Floudas [1], Tawaralani and Sahinidis [15], and the Encyclopedia of Optimization [2].

The major goal of this paper is to show that rigorous error bounds for the optimal value can be computed with little additional computational work in many cases, even for degenerate problems, large (sparse) problems, and problems with uncertain input data. A more detailed presentation of the theory, proofs, algorithms, and several examples can be found in [4].

Independently and at the same time Neumaier and Shcherbina [12] have investigated rigorous error bounds for mixed-integer linear programming problems. In their paper (i) rigorous cuts, (ii) a certificate of infeasibility, and (iii) a rigorous lower bound for linear programming problems with exact input data and finite simple bounds are presented. Our focus is on problems with uncertain input data, and simple bounds which may be infinite. The overlapping part of both papers is the case (iii); a quick look shows that in this case both presented

C. Bliek et al. (Eds.): COCOS 2002, LNCS 2861, pp. 59–70, 2003.
© Springer-Verlag Berlin Heidelberg 2003

bounds coincide. Moreover, [12] contains an innocent looking integer linear programming problem with small integers as input data where many commercial state-of-the-art solvers failed.

2   A Rigorous Upper Bound

We consider the linear programming problem

$$\min_{x \in F} c^T x, \quad F := \{x \in \mathbf{R}^n : Ax \leq a, \ Bx = b, \ \underline{x} \leq x \leq \overline{x}\}, \tag{1}$$

with matrices $A \in \mathbf{R}^{m \times n}$, $B \in \mathbf{R}^{p \times n}$, and vectors $c, x \in \mathbf{R}^n$, $a \in \mathbf{R}^m$, and $b \in \mathbf{R}^p$. F is called the set of *feasible solutions*, and f^* is the *optimal value*. Also infinite simple bounds are allowed; that is $\underline{x}_j := -\infty$ or $\overline{x}_j := +\infty$ for some $j \in \{1, \ldots, n\}$, and hereafter we use the arrangement $0 \cdot (+\infty) = 0 \cdot (-\infty) = 0$.

Some or all input data may be uncertain. We describe uncertainties by considering a $m \times n$ interval matrix $\mathbf{A} \in \mathrm{IR}^{m \times n}$, a $p \times n$ interval matrix $\mathbf{B} \in \mathrm{IR}^{p \times n}$, and interval vectors $\mathbf{a}, \mathbf{b}, \mathbf{c}$. These interval input data define a family of lp-problems $P := (A, B, a, b, c) \in \mathbf{P} := (\mathbf{A}, \mathbf{B}, \mathbf{a}, \mathbf{b}, \mathbf{c})$. To indicate the dependency of the notation above from $P \in \mathbf{P}$, we write sometimes $F(P), f^*(P), x(P)$, etc. In the following only an elementary knowledge of interval arithmetic, including the interval operations between interval matrices and interval vectors, is required; see for example Neumaier [11].

For interval quantities $\mathbf{A}$ we also use the notation $[\underline{A}, \overline{A}]$. Real quantities A are embedded in the interval quantities by identifying $A = \mathbf{A} = [A, A]$, and operations between real and interval quantities are interval operations. For interval quantities $\mathbf{A}, \mathbf{B}$ we define

$$\check{\mathbf{A}} := (\underline{A} + \overline{A})/2 \quad \text{as the } \textit{midpoint}, \tag{2}$$

$$\mathrm{rad}(\mathbf{A}) := (\overline{A} - \underline{A})/2 \quad \text{as the } \textit{radius}, \tag{3}$$

$$|\mathbf{A}| := \sup\{|A| : A \in \mathbf{A}\} \quad \text{as the } \textit{absolute value}, \tag{4}$$

$$\mathbf{A}^+ := \max\{0, \overline{A}\}, \tag{5}$$

$$\mathbf{A}^- := \min\{0, \underline{A}\}. \tag{6}$$

Moreover, the comparison is defined by

$$\mathbf{A} \leq \mathbf{B} \quad \text{iff} \quad \overline{A} \leq \underline{B},$$

and other relations are defined analogously.

The approach for computing a rigorous upper bound is the determination of an interval vector $\mathbf{x}$ which contains a relative interior feasible solution for every lp-problem $P \in \mathbf{P}$. In order to obtain an upper bound close to the optimal value, this interval vector should be at the optimal solution. On the other hand, for the verification of feasibility the interval vector must be sufficiently far away from degeneracy and infeasibility. The next theorem gives favourable characteristics for $\mathbf{x}$.

Theorem 1. *Let* $\mathbf{P} := (\mathbf{A}, \mathbf{B}, \mathbf{a}, \mathbf{b}, \mathbf{c})$ *be a family of lp-problems with input data* $P \in \mathbf{P}$, *and suppose that there exists an interval vector* $\mathbf{x} \in \mathbf{IR}^n$ *such that*

$$\mathbf{A} \cdot \mathbf{x} \leq \mathbf{a}, \ \underline{x} \leq \mathbf{x} \leq \overline{x}, \tag{7}$$

and

$$\forall B \in \mathbf{B}, \ \forall b \in \mathbf{b} \ \exists x \in \mathbf{x}: \ Bx = b. \tag{8}$$

Then for every $P \in \mathbf{P}$ *there exists a primal feasible solution* $x(P) \in \mathbf{x}$, *and the inequality*

$$\sup_{P \in \mathbf{P}} f^*(P) \leq \overline{f}^* := \max(\mathbf{c}^T \cdot \mathbf{x}) \tag{9}$$

is satisfied.

PROOF. Let $P = (A, B, a, b, c) \in \mathbf{P}$ be fixed. The condition (8) implies that there exists a $x(P) \in \mathbf{x}$ with $B \cdot x(P) = b$, and (7) yields that $x(P)$ is a primal feasible solution. Hence, $c(P)^T x(P) \geq f^*(P)$, and (9) follows by inclusion monotonicity. $\square$

For computing an appropriate interval vector $\mathbf{x}$, we compute initially an approximate optimal solution $\tilde{x}$ of a -to the relative interior of $F(\check{\mathbf{P}})$- slightly perturbed problem, and then $n-p$ variables of $\tilde{x}$ are fixed. The condition (8) leads to a $p \times p$ interval linear system. This system is solved by using an appropriate solver for interval linear systems. The computed interval components together with the fixed components provide an interval vector $\mathbf{x}$ satisfying condition (8). If additionally the inequalities (7) hold true, then the right hand side of (9) yields an upper bound. If (7) is not fulfilled, then the problem is perturbed two or three times again.

A similar approach was first proposed by Hansen [3] for proving existence of a feasible point for nonlinear equations -with exact input data- within a bounded box, and was ivestigated and improved numerically by Kearfott [5], [6], [7]. Corresponding algorithms are implemented in his software package GlobSol. A more detailed description of the above algorithm is given in [4], where we have adapted this technique for the linear case, but focussing on uncertain input data and unbounded variables. This causes a special choice of deflation parameters, other fixed variables, and an iterative process.

3   A Rigorous Lower Bound

We consider the convex problem with exact input data

$$\begin{aligned} &\min f(x) \\ &\text{s.t. } G(x) \leq 0 \\ &\quad\quad Bx = b \\ &\quad\quad \underline{x} \leq x \leq \overline{x}, \end{aligned} \tag{10}$$

where $f : \mathbf{R}^n \to \mathbf{R}$, $G : \mathbf{R}^n \to \mathbf{R}^m$, $B \in \mathbf{R}^{p \times n}$, $b \in \mathbf{R}^p$, and f and all components G_i of G are convex functions. The optimal value is denoted by f^*. We denote the gradients of f and G, or in the non-smooth case the subgradients, by $\nabla f(x)$ and $\nabla G(x) = (\nabla G_1(x), \ldots, \nabla G_m(x))$.

Lemma 1. *Let $\tilde{x} \in \mathbf{R}^n$, $\tilde{y} \in \mathbf{R}_+^m$, $\tilde{z} \in \mathbf{R}^p$, and define the defect*

$$d := \nabla f(\tilde{x}) + \nabla G(\tilde{x}) \cdot \tilde{y} - B^T \tilde{z}. \tag{11}$$

Then

$$\underline{f}^* := f(\tilde{x}) + (G(\tilde{x}) - \nabla G(\tilde{x})^T \tilde{x})^T \tilde{y} + b^T \tilde{z} \\ + \underline{x}^T d^+ + \overline{x}^T d^- - \nabla f(\tilde{x})^T \tilde{x} \tag{12}$$

is a lower bound of f^*, that is $\underline{f}^* \leq f^*$.

PROOF. The convexity implies that for each $\tilde{x} \in \mathbf{R}^n$ the inequalities

$$f(x) \geq f(\tilde{x}) + \nabla f(\tilde{x})^T (x - \tilde{x}) \quad \text{for } x \in \mathbf{R}^n, \tag{13}$$

and

$$G(x) \geq G(\tilde{x}) + \nabla G(\tilde{x})^T (x - \tilde{x}) \quad \text{for } x \in \mathbf{R}^n, \tag{14}$$

are satisfied.

The inequalities (13) and (14) provide affine lower bound functions of f and G, and by replacing in (10) the convex functions f and G_i for $i = 1, \dots, m$ by their affine lower bound functions, we obtain a linear programming problem

$$\begin{aligned} \min \; & f(\tilde{x}) + \nabla f(\tilde{x})^T (x - \tilde{x}) \\ \text{s.t. } & G(\tilde{x}) + \nabla G(\tilde{x})^T (x - \tilde{x}) \leq 0 \\ & Bx = b \\ & \underline{x} \leq x \leq \overline{x}, \end{aligned} \tag{15}$$

which has the property that each feasible solution for problem (10) is feasible for (15), and the optimal value of (15) is a lower bound for the optimal value of (10).

This linear program can be written in the form

$$\begin{aligned} \min \; & \nabla f(\tilde{x})^T x + (f(\tilde{x}) - \nabla f(\tilde{x})^T \tilde{x}) \\ \text{s.t. } & \nabla G(\tilde{x})^T x \leq -G(\tilde{x}) + \nabla G(\tilde{x})^T \tilde{x} \\ & Bx = b \\ & \underline{x} \leq x \leq \overline{x}. \end{aligned}$$

The corresponding dual problem is

$$\begin{aligned} \max \; & (G(\tilde{x}) - \nabla G(\tilde{x})^T \tilde{x})^T y + b^T z + \\ & \underline{x}^T u - \overline{x}^T v + f(\tilde{x}) - \nabla f(\tilde{x})^T \tilde{x} \\ \text{s.t. } & -\nabla G(\tilde{x}) y + B^T z + u - v = \nabla f(\tilde{x}) \\ & y \geq 0, \; u \geq 0, \; v \geq 0. \end{aligned} \tag{16}$$

If we define

$$\tilde{u} := d^+, \quad \tilde{v} := -d^-,$$

then $\tilde{u} \geq 0$, $\tilde{v} \geq 0$, and from $\tilde{y} \geq 0$ we obtain immediately that $\tilde{y}, \tilde{z}, \tilde{u}, \tilde{v}$ are feasible for the dual problem with the objective value $\underline{f}^*$, which is by duality theory less than or equal f^*. $\qquad \square$

Using this lemma the following theorem can be proved which allows the computation of a rigorous lower error bound for linear programming problems with interval input data.

Theorem 2. *Let* $\mathbf{P} := (\mathbf{A}, \mathbf{B}, \mathbf{a}, \mathbf{b}, \mathbf{c})$ *be a family of lp-problems with input data* $P \in \mathbf{P}$. *Suppose that* $\tilde{y} \in \mathbf{R}_+^m$, $\tilde{z} \in \mathbf{R}^p$, *and let*

$$\mathbf{d} := \mathbf{c} + \mathbf{A}^T \cdot \tilde{y} - \mathbf{B}^T \cdot \tilde{z}. \tag{17}$$

Then the inequality

$$\inf_{P \in \mathbf{P}} f^*(P) \geq \underline{f}^* := \min(-\mathbf{a}^T \cdot \tilde{y} + \mathbf{b}^T \cdot \tilde{z} + \underline{x}^T \cdot \mathbf{d}^+ + \overline{x}^T \cdot \mathbf{d}^-) \tag{18}$$

is fulfilled.

PROOF. Let $P = (A, B, a, b, c) \in \mathbf{P}$ be a fixed real linear programming problem, and let

$$G(x) := Ax - a \quad \text{and} \quad f(x) := c^T x.$$

Then (11) yields

$$d = c + A^T \tilde{y} - B^T \tilde{z}$$

and from (12) it follows that

$$\begin{aligned} \underline{f}^*(P) &= c^T \tilde{x} + (A\tilde{x} - a - A\tilde{x})^T \tilde{y} + b^T \tilde{z} + \underline{x}^T d^+ + \overline{x}^T d^- - c^T \tilde{x} \\ &= -a^T \tilde{y} + b^T \tilde{z} + \underline{x}^T d^+ + \overline{x}^T d^-. \end{aligned}$$

Inclusion monotonicity yields the inequality (18). □

In the case where all simple bounds are finite, the right hand side of (18) provides a finite lower bound, and the following postprocessing algorithm, consisting of three simple steps, suffice:

We assume that an lp-solver has already computed an approximate optimal solution $\tilde{x}$ with corresponding Lagrange multipliers $\tilde{y}, \tilde{z}$ of the midpoint problem of $\mathbf{P}$.

(1) Set $\tilde{y} := \tilde{y}^+$.
(2) Compute $\mathbf{d} := \mathbf{c} + \mathbf{A}^T \tilde{y} - \mathbf{B}^T \tilde{z}$.
(3) Compute $\underline{f}^* := \min(-\mathbf{a}^T \cdot \tilde{y} + \mathbf{b}^T \cdot \tilde{z} + \underline{x}^T \cdot \mathbf{d}^+ + \overline{x}^T \cdot \mathbf{d}^-)$.

Nothing is assumed about the quality of these approximations. However, this lower bound depends mainly on the accuracy of the lp-solver, and the radius of the interval input data. It can be proved that in case of exact input data, and an exactly computed optimal solution with corresponding exact Lagrange multipliers, the lower bound coincides with the optimal value. Moreover, we point out that no overestimation occurs in the calculation of the defect $\mathbf{d}$, because the interval matrices $\mathbf{A}$ and $\mathbf{B}$ are multiplied by real vectors.

This algorithm requires only few interval matrix-vector operations, and is practicable also for large-scale problems, since appropriately implemented matrix-vector computations exploit sparse structures. Degenerate problems (that is, more than n constraints are active) can be rigorously bounded from below without any difficulties, provided the lp-solver has computed sufficiently good approximations.

If some of the variables are unbounded, then the terms $\underline{x}^T \mathbf{d}^+$ or $\overline{x}^T \mathbf{d}^-$ in (18) are infinite iff there exists an index j with $\underline{x}_j = -\infty$ and $\mathbf{d}_j^+ > 0$ or $\overline{x}_j = +\infty$ and $\mathbf{d}_j^- < 0$, respectively, yielding an infinite rigorous lower bound $\underline{f}^*$. In order to obtain a finite lower bound $\underline{f}^*$ as well for unbounded variables, the idea is to solve a perturbed linear programming problem which produces approximations $\tilde{y}$ and $\tilde{z}$ such that

$$\mathbf{d}_j^+ = 0 \text{ for } \underline{x}_j = -\infty, \text{ and } \mathbf{d}_j^- = 0 \text{ for } \overline{x}_j = +\infty. \tag{19}$$

If all variables are bounded at least on one side then we try to achieve this goal by perturbing the midpoint of $\mathbf{c}$ componentwise by

$$c_j(\varepsilon) := \begin{cases} \check{c}_j + \varepsilon_j, & \text{if } \underline{x}_j = -\infty \\ \check{c}_j - \varepsilon_j, & \text{if } \overline{x}_j = +\infty \\ \check{c}_j & \text{otherwise,} \end{cases} \tag{20}$$

with appropriately defined $\varepsilon_j > 0$. If ε_j is chosen very large, the computed approximations of the perturbed problem $\tilde{y}$ and $\tilde{z}$ are far away from the optimal solution of the midpoint problem, yielding an unnecessary overestimation of the lower bound $\underline{f}^*$. On the other hand, too small ε_j may imply an infinite lower bound, because condition (19) is not satisfied. We choose

$$\begin{aligned} \varepsilon_j = \; &2((\text{rad}(\mathbf{c}_j) + \text{rad}(\mathbf{A}_{:j})^T |\tilde{y}| + \text{rad}(\mathbf{B}_{:j})|\tilde{z}|) \\ &+\varepsilon_a(|c_j| + |(A_{:j})^T||\tilde{y}| + |(B_{:j})^T||\tilde{z}|)), \end{aligned} \tag{21}$$

for j with $\overline{x}_j = +\infty$ or $\underline{x}_j = -\infty$, where ε_a is the adjusted accuracy of the lp-solver, and $\tilde{y}, \tilde{z}$ are approximate dual optimal solutions (i.e. Lagrange multipliers corresponding to $Ax \le a$ and $Bx = b$, respectively). Hence, ε_j takes into consideration the accuracy and the radii of the input data.

The algorithm for variables which are unbounded at most on one side is iterative and consists of the following steps:

Set ε_j as defined in formula (21) for the unbounded variables, and set $k = 0$.
Until $k > k_0$

(1) Compute an approximate dual solution $\tilde{y}, \tilde{z}$ of

$$P(\varepsilon) := (\check{\mathbf{A}}, \check{\mathbf{B}}, \check{\mathbf{a}}, \check{\mathbf{b}}, c(\varepsilon))$$

where $c(\varepsilon)$ is defined by (20).
If the lp-solver cannot find an approximate solution (i.e. the solver displays that the dual is infeasible), then STOP: *Lower bound is minus infinity.*

(2) Redefine $\tilde{y}$ by setting all negative components equal to zero (i.e. $\tilde{y} \in \mathbf{R}_+^m$).

(3) Compute $\mathbf{d} := \mathbf{c} + \mathbf{A}^T \tilde{y} - \mathbf{B}^T \tilde{z}$.

(4) If (19) is satisfied, then STOP: *The rigorous lower bound is equal to*

$$\underline{f}^* = \min(-\mathbf{a}^T \tilde{y} + \mathbf{b}^T \tilde{z} + \underline{x}^T \mathbf{d}^+ + \overline{x}^T \mathbf{d}^-).$$

(5) Set $\varepsilon_j = 2 \cdot \varepsilon_j$ for the unbounded variables.
(6) Set $k = k + 1$.

The number k_0 is the maximal number of iterations. In our present implementation we have chosen $k_0 = 5$, but very frequently at most three iterations suffice.

For the variables x_j which are unbounded on both sides condition (19) leads to $\mathbf{d}_j = 0$. This means that for these indices the interval linear equations

$$\forall A \in \mathbf{A}, B \in \mathbf{B}, c \in \mathbf{c} \; \exists y, z :$$
$$-(A_{\cdot j})^T y + (B_{\cdot j})^T z = c_j \tag{22}$$

must be solved. Here, we can use the same technique as for the interval linear equations in the case of the upper bound.

Several modifications of this algorithm are possible and may yield improvements at least for special classes of problems. For example, in cases where some of the simple bounds are small and some are very large, it is advantageous to take the second algorithm and to choose the perturbed problem by defining

$$c_j(\varepsilon) := \begin{cases} \check{c}_j + \varepsilon_j, \text{ if } |\underline{x}_j| \geq |\overline{x}_j| \\ \check{c}_j - \varepsilon_j, \text{ if } |\underline{x}_j| < |\overline{x}_j| \end{cases} \tag{23}$$

for components with large $|\underline{x}_j|$ or $|\overline{x}_j|$.

4   Examples

Example 1. For the purpose of illustration we consider a degenerate lp-problem, where the constraints are taken from Vanderbei [16], page 36

$$\begin{aligned} \min - x_1 \; - \; & x_2 - 4x_3 \text{ subject to} \\ x_1 \quad\quad & + 2x_3 \leq 2 \\ & x_2 + 2x_3 \leq 2 \\ 0 \leq x_1 \quad , x_2 \quad & , x_3 \leq 100. \end{aligned} \tag{24}$$

This problem is illustrated in the following figure.

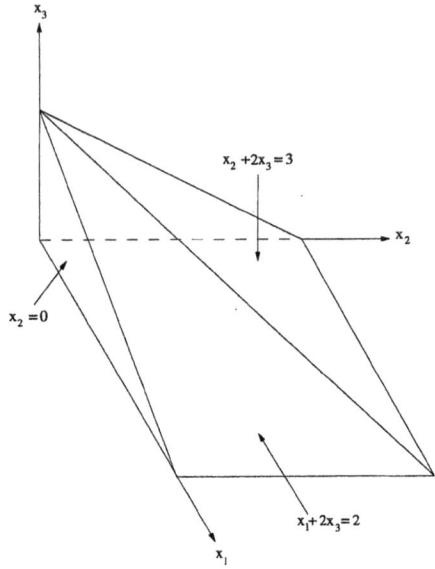

Obviously, $x^* = (0, 0, 1)^T$ is optimal with value $f^* = -4$. The vector x^* is the intersection of four of the facets, and not of three facets as one would normally expect, that is x^* is degenerate. Moreover, it is not a simple degeneracy caused by redundant constraints, since deleting only one of the above constraints changes the set of feasible solutions. The dual problem has no unique optimal solution; one of the optimal solutions has Lagrange multipliers $y^* = (1, 1)$ corresponding to the non-simple inequalities.

Now, we want to illustrate the behaviour of the error bounds for interval problems. To simplify matters we only consider the effect of interval input data, but not of rounding errors. Each nonzero coefficient of the objective function and the constraints is replaced by an interval with radius r where $0 \leq r \leq 1/3$, yielding an interval problem $\mathbf{P}$, where $\mathbf{B}$ and $\mathbf{b}$ are empty, and

$$\mathbf{A} = \begin{pmatrix} [1 - r, 1 + r] & 0 & [2 - r, 2 + r] \\ 0 & [1 - r, 1 + r] & [2 - r, 2 + r] \end{pmatrix}$$

$$\mathbf{a} = ([2 - r, 2 + r], \quad [2 - r, 2 + r])^T, and$$

$$\mathbf{c} = ([-1 - r, -1 + r], \quad [-1 - r, -1 + r], \quad [-4 - r, -4 + r])^T.$$

Let $\mathbf{x} = \tilde{x} = (0, 0, 1 - \frac{3}{2}r)^T$ be an initially computed solution in the relative interior of the primal feasible set.

A short computation shows that

$$\mathbf{A} \cdot \mathbf{x} = \begin{pmatrix} [2 - 4r + \frac{3}{2}r^2, 2 - 2r - \frac{3}{2}r^2] \\ [2 - 4r + \frac{3}{2}r^2, 2 - 2r - \frac{3}{2}r^2] \end{pmatrix} \leq \mathbf{a}$$

and $0 \leq \mathbf{x} \leq 100$. Hence, all assumptions of Theorem 1 are satisfied, and

$$\sup_{p \in \mathbf{P}} f^*(P) \le \overline{f}^* = \max([-4 - r, -4 + r] \cdot (1 - \frac{3}{2}r)) = -4 + 7r - \frac{3}{2}r^2.$$

For the computation of the lower bound we use the first algorithm in Section 3, and assume that the lp-solver has calculated the optimal dual solution $\tilde{y} = y^* = (1, 1)^T$. Then the defect is

$$\mathbf{d} = \mathbf{c} + \mathbf{A}^T \tilde{y} = \begin{pmatrix} [-1 - r, -1 + r] \\ [-1 - r, -1 + r] \\ [-4 - r, -4 + r] \end{pmatrix} + \begin{pmatrix} [1 - r, 1 + r] & 0 \\ 0 & [1 - r, 1 + r] \\ [2 - r, 2 + r] & [2 - r, 2 + r] \end{pmatrix} \begin{pmatrix} 1 \\ 1 \end{pmatrix}$$

$$= \begin{pmatrix} [-2r, 2r] \\ [-2r, 2r] \\ [-3r, 3r] \end{pmatrix},$$

and the lower bound is equal to

$$\underline{f}^* = \min(-[2 - r, 2 + r] \cdot 1 - [2 - r, 2 + r] \cdot 1 +$$

$$\begin{pmatrix} 0 \\ 0 \\ 0 \end{pmatrix}^T \cdot \begin{pmatrix} [0, 2r] \\ [0, 2r] \\ [0, 3r] \end{pmatrix} + \begin{pmatrix} 100 \\ 100 \\ 100 \end{pmatrix}^T \begin{pmatrix} [-2r, 0] \\ [-2r, 0] \\ [-3r, 0] \end{pmatrix})$$

$$= -4 - 702r.$$

Although we obtain the optimal value $\underline{f}^* = f^*$ for radius $r = 0$, the lower bound overestimates the optimal value. The reason for this large overestimation is the nonzero vector $\mathbf{d}^-$ which is multiplied with the large simple bounds $\overline{x}$. This overestimation can be reduced by using the second algorithm of Section 3, and solving the perturbed midpoint problem. Formula (21) yields $\varepsilon_j = 4r$ for $j = 1, 2$ and $\varepsilon_3 = 6r$. After the first iteration condition (19) is not satisfied. But in the second iteration formula (23) implies

$$c_j(\varepsilon) = \begin{cases} -1 - 8r & , j = 1, 2 \\ -4 - 12r & , j = 3 \end{cases}$$

The dual optimal solution of this perturbed problem is $\tilde{y} = (1 + 3r, 1 + 3r)^T$ yielding the optimal value

$$f^*(\varepsilon) = -a^T \cdot \tilde{y} = -4 - 12r = c(\varepsilon)^T \cdot x^*.$$

For the lower bound, we compute the defect

$$\mathbf{d} = \mathbf{c} + \mathbf{A}^T \tilde{y} = \begin{pmatrix} [-1 - r, -1 + r] \\ [-1 - r, -1 + r] \\ [-4 - r, -4 + r] \end{pmatrix} + \begin{pmatrix} [1 - r, 1 + r] & 0 \\ 0 & [1 - r, 1 + r] \\ [2 - r, 2 + r] & [2 - r, 2 + r] \end{pmatrix} \begin{pmatrix} 1 + 3r \\ 1 + 3r \end{pmatrix}$$

$$= \begin{pmatrix} [r - 3r^2, 5r + 3r^2] \\ [r - 3r^2, 5r + 3r^2] \\ [9r - 6r^2, 15r + 6r^2] \end{pmatrix}.$$

Since $\tilde{y} \geq 0$, $\underline{x} = 0$ and $\mathbf{d} \geq 0$ for $0 \leq r \leq \frac{1}{3}$ we obtain $\mathbf{d}^- = 0$, and the terms $\underline{x}^T \mathbf{d}^+$ and $\overline{x}^T \mathbf{d}^-$ vanish, yielding the lower bound of the optimal value

$$\underline{f}^* = \min(-\,[2-r, 2+r] \cdot (1+3r) - [2-r, 2+r] \cdot (1+3r)) = -4 - 14r - 6r^2.$$

Now the overestimation is drastically reduced.

Example 2. We consider random test problems which are generated with regard to the construction of Rosen and Suzuki [14]. The primal solution x^* is constructed by uniformly distributing 90% of the components in the interval $(0,1)$, and the remaining components are set equal to zero. The lower and upper bounds $\underline{x}_i$ and $\overline{x}_i$ are set equal to zero and one for every component x_i, respectively. The dual vector y^* is generated by uniformly distributing $n - p$ components in $(-1,0)$ while the remaining components are set equal to zero. The coefficients of z^*, A and B are uniformly distributed in $(-1,1)$. The right hand side a is chosen such that the first $n - p$ inequalities are active, and the right hand side $b := Bx^*$. The coefficients of the objective function are generated by the equation $c := A^T y^* + B^T z^*$. This construction ensures that these test problems are degenerate with known optimum; that is, x^* is primal optimal and $(y^*, z^*, 0, 0)$ is dual optimal. Moreover, we consider interval input data with $\operatorname{rad}(\mathbf{P}) = r \cdot \check{\mathbf{P}}$, where $\check{\mathbf{P}}$ is the previously defined random midpoint problem.

The following numerical results are obtained by using the linear programming routine *lp-solve 4.01* [9], and the interval arithmetic package PROFIL/BIAS [8]. The adjusted accuracy of *lp-solve 4.01* is $\varepsilon_a = 10^{-8}$. I wish to thank Christian Keil for the implementation of the error bounds.

In Tables 1 and 2 and we display the number of inequalities m, the number of variables n, the number of equations p, the relative error $|\overline{f}^* - f^*|/|\underline{f}^*|$ of the rigorous bounds, and the ratios $t_{\underline{f}^*}/t_s$, $t_{\overline{f}^*}/t_s$ where t_s denotes the time required by *lp-solve* applied to the midpoint problem, $t_{\underline{f}^*}$ is the time for the lower bound, and $t_{\overline{f}^*}$ denotes the time required by the upper bound.

For this test set it can be seen that the rigorous lower and upper bounds can be computed very cheap for small radius r compared with the time required to

Table 1. Results for randomly generated problems $r := 0$

m	n	p	$\lvert\overline{f}^* - f^*\rvert/\lvert\underline{f}^*\rvert$	$t_{\underline{f}^*}/t_s$	$t_{\overline{f}^*}/t_s$
250	100	0	$4.2038857035e-06$	0.035	0.039
500	100	0	$1.7934507622e-06$	0.028	0.032
1000	250	0	$1.5652711068e-05$	0.010	0.010
0	500	100	$1.3705844636e-07$	0.006	0.017
0	1000	250	$2.7512815545e-08$	0.003	0.010
0	5000	500	$2.9727288243e-08$	0.001	0.002
100	50	25	$1.8199522305e-06$	0.071	0.148

Table 2. Results for randomly generated problems with $r := 10^{-8}$

m	n	p	$|\tilde{f}^* - f^*|/|f^*|$	$t_{\underline{f}^*}/t_s$	$t_{\overline{f^*}}/t_s$
250	100	0	1.2173724773e-05	0.034	0.042
500	100	0	5.1655285808e-06	0.028	0.033
1000	250	0	4.4784287608e-05	0.010	0.010
0	500	100	4.8616339428e-05	0.006	0.017
0	1000	250	1.9379754477e-04	0.003	0.010
0	5000	500	6.4684160307e-04	0.001	0.002
100	50	25	2.9092666980e-06	0.071	0.214

solve the problem. For increasing radius the upper requires more time and may be not computable.

We have investigated further numerical examples. Summarizing, the main observations, which are also supported by our theoretical analysis, can be roughly described as follows:

- The lower bound can be computed efficiently and is very sharp for linear programming problems with (i) input data which are exact or of very small radius, (ii) moderate finite simple bounds, and (iii) an approximate optimal solution of good accuracy computed by the lp-solver. This is caused by the computed defect which is almost equal to the exact defect yielding at most a small overestimation. Moreover, only few operations (compared with the lp-solver) are necessary.

- The same properties has the lower bound in the case of degenerate problems. If additionally the radius and the simple bounds increase, the computational work for the lower bound is not influenced, but a large overestimation may occur. This overestimation can be significantly reduced by using the second algorithm of Section 3, as demonstrated in Example 1. But this latter algorithm is more expensive.

- If the primal or dual problem is ill-posed (i.e. the distance to primal infeasibility or dual infeasibility is zero) but (i) and (ii) are satisfied, then a reasonable finite lower bound can be obtained with formula (18) provided the lp-solver has computed a sufficiently good approximate solution.

- A finite lower bound cannot be computed for unbounded variables whereby the corresponding sign-conditions (19) cannot be fulfilled. This occurs if the dual inequalities corresponding to the unbounded variables cannot be strictly solved, which implies that the dual problem is ill-posed.

- The upper bound additionally proves the existence of primal feasible solutions for all lp-problems of the family **P** (see Theorem 1). Hence, this bound is more restricted, and can be used only for problems which have non-empty relative interior.

5 Conclusions

We have presented rigorous error bounds for linear programming problems with uncertain input data. These bounds can be computed very cheap in many cases, and require only approximations calculated by lp-solvers. Beside the property to handle with lp-solvers in a safe manner, the bounds can also be used in branch-and-bound methods for rigorously solving global optimization and mixed integer problems, whenever linear relaxations must be solved.

References

[1] C. A. Floudas. *Deterministic Global Optimization - Theory, Methods and Applications*, volume 37 of *Nonconvex Optimization and Its Applications*. Kluwer Academic Publishers, Dordrecht, Boston, London, 2000.

[2] C. A. Floudas and P. M. Pardalos. *Encyclopedia of Optimization*. Kluwer Academic Publishers, 2001.

[3] E. R. Hansen. *Global Optimization using Interval Analysis*. Marcel Dekker, New York, 1992.

[4] C. Jansson. Rigorous Lower and Upper Bounds in Linear Programming. Technical Report 02.1, Forschungsschwerpunkt Informations- und Kommunikationstechnik, TU Hamburg-Harburg, 2002.
http://www.ti3.tu-harburg.de/paper/jansson/verification.ps.

[5] R. B. Kearfott. On proving existence of feasible points in equality constrained optimization problems. Preprint, Deptartment of Mathematics, Univ. of Southwestern Louisiana, U. S. L. Box 4-1010, Lafayette, La 70504, 1994.

[6] R. B. Kearfott. On verifying feasiblility in equality constrained optimization problems. Technical report, Deptartment of Mathematics, Univ. of Southwestern Louisiana, 1994.

[7] R. B. Kearfott. *Rigorous Global Search: Continuous Problems*. Kluwer Academic Publisher, Dordrecht, 1996.

[8] O. Knüppel. PROFIL / BIAS — A Fast Interval Library. *Computing*, 53:277–287, 1994.

[9] lp-solve 4.0-1. http://packages.debian.org/unstable/math/lp-solve.html.

[10] NETLIB Linear Programming Library. http://www.netlib.org/lp.

[11] A. Neumaier. *Introduction to Numerical Analysis*. Cambridge University Press, 2001.

[12] A. Neumaier and O. Shcherbina. Safe bounds in linear and mixed-integer programming. Submitted for publication.

[13] F. Ordóñez and R. M. Freund. Computational experience and the explanatory value of condition numbers for linear optimization. Technical Report MIT Operations Research Center Working Paper OR361-02, MIT, 2002. submitted for publication to SIAM J. Optimization.
http://web.mit.edu/rfreund/www/CVfreund.pdf.

[14] J. B. Rosen and S. Suzuki. Construction of Nonlinear Programming Test Problems. *Communication of ACM*, 8:113, 1965.

[15] M. Tawaralani and N. V. Sahinidis. *Convexification and Global Optimization in Continuous and Mixed-Integer Nonlinear Programming*. Kluwer Academic Publishers, 2002.

[16] R. J. Vanderbei. *Linear Programming: Foundations and Extensions*. Kluwer Academic Publishers, 1996.

Minimal and Maximal Real Roots
of Parametric Polynomials
Using Interval Analysis*

The COPRIN Project

INRIA
BP 93, 06902 Sophia-Antipolis, France
Phone: 33 4 92 38 77 61
jean-pierre.merlet@sophia.inria.fr
http://www-sop.inria.fr/coprin/welcome.html

Abstract. In this paper we are interested in parametric polynomials i.e. polynomials whose coefficients are analytical functions of parameters that are restricted to lie within given ranges. We propose algorithms based on interval analysis to solve various problems such as finding the extremal real roots for the set of polynomials and determining an approximation of the set of parameters values such that all the polynomials have their root real part in a given range. Realistic application examples are presented in the field of robotics and control theory.

Keywords: parametric polynomial, eigenvalues, interval analysis

1 Notation

We use in this paper classical notation in interval analysis. The lower and upper bound of an interval a will be denoted $\underline{a}$, $\overline{a}$. The *width* of an interval a is the difference $\overline{a} - \underline{a}$.

2 Motivation

To introduce the problems that we intend to address in this paper we will start with an application example in robotics. The kinematics of a robot may be defined as the relationship between:

- the position/orientation of the end-effector X, i.e. the location of a rigid body of the robot that is instrumented with tools (e.g a gripper) for performing various tasks
- the joint variables Θ: the values of these variables are measured with a set of sensors and may be changed by using actuators

---

* Corresponding author: J-P. Merlet

C. Bliek et al. (Eds.): COCOS 2002, LNCS 2861, pp. 71–86, 2003.

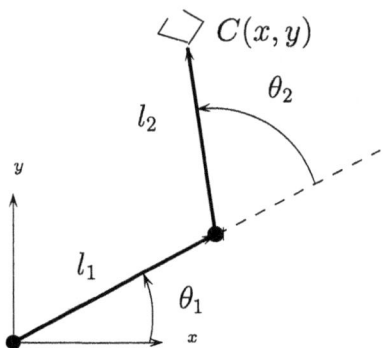

Fig. 1. A 2 d.o.f. robot

and may be written as $F(X, \Theta, \mathcal{P}) = 0$, where $\mathcal{P}$ is the set of parameters that define the geometry of the robot. The determination of this relation is essential for a robot as the motion of the end-effector is obtained by an appropriate control of the actuated joints.

The measurements of the joint variables are not perfect: to each sensor is associated a known error and these errors are represented by an error vector $\Delta\Theta$. These errors induce an error ΔX on the positioning of the end-effector. Using a first-order approximation we have:

$$\Delta X = J(X, \mathcal{P}, \Theta)\Delta\Theta \qquad (1)$$

where J is called the *jacobian matrix of the robot*. This matrix is the product of the jacobian of F by a non singular constant matrix and is a function of $\mathcal{P}$, the location X of the gripper or/and the values of the joint coordinates Θ.

For example if we consider a 2 degrees of freedom (d.o.f.) planar robot (figure 1) the location of the end-effector is defined by the coordinates x, y of its center C, while the joint variables are θ_1, θ_2

We have:

$$x = l_1 \cos\theta 1 + l_2 \cos(\theta_2 - \theta_1)$$
$$y = l_1 \sin\theta 1 + l_2 \sin(\theta_2 - \theta_1)$$

Hence:

$$\begin{pmatrix} \Delta x \\ \Delta y \end{pmatrix} = \begin{pmatrix} l_1 \sin\theta_1 - l\sin(\theta_1 - \theta_2) & l_2 \sin(\theta_1 - \theta_2) \\ l_1 \cos\theta_1 - l_2 \cos(\theta_1 - \theta_2) & l_2 \cos(\theta_1 - \theta_2) \end{pmatrix} \begin{pmatrix} \Delta\theta_1 \\ \Delta\theta_2 \end{pmatrix}$$

This relation is essential to study the effect of the sensor measurement errors on the accuracy of the positioning of the platform.

3 Accuracy and Joint Errors

The errors on the joint measurements are individually *bounded* and as equation (1) is linear we may assume:

$$||\Delta\Theta|| \leq 1 \Longleftrightarrow |\Delta\Theta_i| \leq 1 \quad \forall i \tag{2}$$

The use of the L_∞ norm indicates that the error bound on one sensor is independent from the error bounds on the others sensors. As $\Delta\Theta = J^{-1}\Delta X$ we get:

$$||\Delta X^T J^{-T} J^{-1} \Delta X|| \leq 1$$

In other words we have a linear mapping between the joints errors and the end-effector positioning errors. In the joint error space equation (2) defines an hypercube which is mapped into a convex polyhedron in the end-effector error space. The maximal and minimal eigenvalues of $J^{-T}J^{-1}$ give some information on the largest and lowest positioning errors for the end-effector. Note that $J^{-T}J^{-1}$ is a symmetrical, positive definite matrix and hence that all the eigenvalues are real and positive.

The maximal and minimal eigenvalues of $J^T J$ are a good indicator for measuring the quality of the design: indeed these eigenvalues indicate the amplification factor between the joints measurement errors and the end-effector positioning errors (note that if one eigenvalue is 0 the robot is in a singular configuration: the end-effector may move without any change of the joint variables). Ideally the minimal and maximal eigenvalues should be identical which will ensure the same behavior of the robot in any motion direction.

In this paper we will consider parametric polynomials that are the characteristic polynomial of a given parametric matrix whose components are arbitrary functions of parameters. A research in the literature on the determination of the eigenvalues of a parametric matrix with arbitrary functions for its components has provided sparse results although we will use some general results of matrix computation [4]. Symbolic algorithms have been proposed for problems in which the polynomial coefficients are polynomials [12]. Similar problems have been addressed in robust control theory [1], a problem that has already been addressed using interval analysis [7]. But the proposed approaches do not allow, in general, to deal with the general case where the coefficients are arbitrary functions of the parameters or to solve all the problems that may be associated to parametric polynomials.

4 Problems to Solve

After this introduction we may define a set of problems that may be interesting to solve:

- find the minimal and maximal eigenvalues of $J^T J$ when the parameter vector X lies within a defined parameter space $\mathcal{W}$: this will allow, for example, to characterize the behavior of the robot over its workspace (i.e. all the poses that the robot end-effector is supposed to reach)

- find the region $\mathcal{R}$ included in a given parameter space $\mathcal{W}$ such that for any point in $\mathcal{R}$ the minimal and maximal eigenvalues lie within a defined range: this will allow, for example, to determine in which part of its workspace the robot will behave efficiently. In control theory the parameters will represent the coefficients of a corrector that allows to robustly control a given system provide that the real part of all the roots of a polynomial are negative. The usual approach allows to determine one possible set of coefficients but we intend here to provide an approximation of the set of all possible coefficients. A specific version of this problem is obtained when the parameter set $\{x_1, x_2, \ldots, x_n\}$ is split into 2 parts $S_1 = \{x_1, x_2, \ldots, x_m\}$ and $S_2 = \{x_{m+1}, x_{m+2}, \ldots, x_n\}$, the problem being to determine all the possible values of the parameters in S_1 such that for any value of the parameters in the second set, within some pre-defined ranges, the roots of the corresponding polynomial lie within the range (see example).
- find the largest square box enclosed in a given parameter space $\mathcal{W}$ such that for any point in the box the minimal and maximal eigenvalues lie within a defined range

There are many difficulties when considering these problems:

- the degree of the polynomial may be relatively high: this implies that the roots may have no analytical expression (and anyway for polynomials with degree larger than 2 analytical expressions for the roots are badly conditioned numerically),
- the parameter space $\mathcal{W}$ may be defined by a set of constraints (e.g. for the the 2D planar robot the set of X such that the θ_1, θ_2 angles lie in given ranges),
- in some cases an analytical form of the characteristic polynomial may be difficult to compute

5 Maximal Real Root of a Characteristic Polynomial

Let P be a polynomial with interval coefficients:

$$P(z) = a_n z^n + a_{n-1} z^{n-1} + \ldots a_0$$

In others words P may be considered as a set of polynomials. If our aim is to use interval analysis-based methods to determine what are the minimal and maximal values of the real roots of any polynomial in the set P the first problem we have to solve is to determine what are the bounds for these values.

5.1 Initial Bound for the Roots

There are numerous methods in classical algebraic geometry to determine bounds on the norm of the roots of a polynomial with fixed value coefficients [2]. Many of these methods may be adapted to deal with polynomials having interval coefficients.

For example one of the Cauchy's theorems states that for a polynomial with fixed-value coefficients if we define A as $A = \text{Max}\{|a_{n-1}|, |a_{n-2}|, \ldots, |a_0|\}$, then any root λ of the polynomial satisfies

$$|\lambda| < 1 + \frac{A}{|a_n|} \tag{3}$$

It may be seen that Cauchy's theorem may also be used for polynomial with interval coefficients provided that the interval for a_n does not contain 0. Indeed the absolute value of an interval a may be calculated as the interval:

$$|a| = [\text{Min}(|\underline{a}|, |\overline{a}|) \text{ if } \underline{a} > 0 \text{ or } \overline{a} < 0, \ 0 \text{ otherwise}, \text{Max}(|\underline{a}|, |\overline{a}|)]$$

Hence, provided that the interval a_n does not contain 0, we may calculate an interval evaluation of the right hand side term of (3). The upper bound of this interval will be an upper bound for the root of the polynomial.

We have extended most of the classical root bounding theorem to the case of interval coefficients and we have developed a numerical procedure $\texttt{MaxRoot}(P)$ that uses these extensions to provide the best bound for the real roots of a polynomial P.

As the polynomial we are considering may be the characteristic polynomial of a matrix we may also use other methods that give bounds on the eigenvalues z of a matrix.

For example if P is the characteristic polynomial of a matrix $A = ((a_{ij}))$, we use the method of the Gershgorin circles [6]. These circles C_i are defined by $|z - a_{ii}| \leq \sum |a_{ji}|, \ \ j \neq i$ and the Gershgorin theorem states that all the eigenvalues are enclosed in the union of the C_i. Furthermore any circle that has no intersection with the other circles contains exactly one eigenvalue and if all the circles are distinct, then each one contains exactly one eigenvalue of the matrix. Similar results hold for the Cassini oval [1]. Here again all these methods may easily be extended to deal with a matrix with interval components.

5.2 Maximal Root Value Tests

As our approach is based on interval analysis we will use sets of intervals called *boxes*. If the coefficients of P are functions of a set of n parameters $x_1, x_2, \ldots, x_n$ while the unknown of the polynomial is z, a box will be a set of intervals for $\{z, x_1, x_2, \ldots, x_n\}$

The purpose of the maximal root value test is to determine if a given box may include values for the parameters such that the corresponding polynomial has a root larger than a given value M (we will assume that for each box we have $\underline{z} \geq M$). The test will return *false* if this is not the case.

The simplest maximal root value test is clearly to compute the interval evaluation of the polynomial $P(z)$: if this evaluation does not contain 0 then the check will return *false*. Note that we use the derivative of the polynomial and eventually the derivative of the coefficients to improve the evaluation of the polynomial. Note that the $\texttt{MaxRoot}$ procedure may also be used for the maximal root value test.

A more sophisticated maximal root value test is based on the Budan-Fourier method for the estimation of the number of roots of the polynomial in the range $[M, \infty[$ [14]. As this method uses only the sign of the evaluation of the polynomial and of its derivative with respect to z at a point it can easily be extended to interval coefficients. The result is an integer interval $[n, m]$ where n and m are the minimal and maximal number of roots up to a power of 2. Hence an interval [2,3] means that the minimal and maximal number of real roots may be described by one of the following set: [0,1], [1,3], [2,3].

Budan-Fourier is, in general, more effective than using the Sturm method [9] since the computation of the functions in the Sturm sequence involves a large number of interval computations.

Another maximal root value test is done using the Kharitonov polynomials [11]. Kharitonov theorem states that all the polynomials in P have positive (negative) roots iff the 4 Kharitonov polynomials have positive (negative) roots, where the Kharitonov polynomials are selected polynomials in P with real-valued coefficients obtained as upper or lower bound of the interval coefficients.

Now assume that we want to determine if the roots of all the polynomials in P are all greater (lower) than a given a. We substitute z in the polynomial by $y + a$ where y is a new variable and compute the coefficients in y of the polynomials $P^\star(y) = P(y + a)$. Using Kharitonov polynomials we may verify if all the roots of $P^\star$ are greater (lower) than 0, in which case all the roots of P will be greater (lower) than a. The main problem here is to calculate efficiently the coefficients of $P^\star$ as described later on.

We use also extensively another method based on the Routh table [1]. The real part of the root of the polynomial will be negative iff the elements of the first column of the table have same sign (the number of roots with positive real part is equal to the number of sign changes in the sequence of the first column of the table). Hence a approach similar to the one used for the Kharitonov polynomials may be used, for the same problem, i.e. determine efficiently the value of the elements of the Routh table for the polynomial $P^\star(y) = P(y + a)$.

Note also that the maximal root value test may be used to *filter* the boxes i.e. to reduce their widths. For example we use in our algorithms the 2B and 3B methods described in [3] (see next section for an example of the use of the 2B method).

5.3   Update of the Maximal Root

If the maximal root value test does not return *false* we will bisect the box. But before doing the bisection we may determine a new value for the current estimate of the maximal root M. For that purpose we compute numerically the roots of the polynomials obtained for some specific points in the box (or alternatively the eigenvalues of the matrix) and compare them with the current estimate. If M is improved a local search strategy is used to refine the bound.

6   Algorithm

We will now describe an algorithm which aims to determine the maximal root of a parametric polynomial, up to a pre-defined accuracy ϵ.

The structure of the algorithm is classical with a list $\mathcal{L}$ of n boxes B_j, indexed by i:

1. initialization of the range for $z=[\underline{Z}, \overline{Z}]$ using the MaxRoot procedure,
2. creation of the first box, $i = 1$, $\mathcal{L} = \{B_1\}$, $n = 1$
3. if $i > n$, then M is the maximal root, exit
4. if check $P(B_i, [M + \epsilon, \overline{Z}])=false$, then $i = i + 1$, goto 3
5. update of M
6. select a variable of the box, bisect B_i, add the 2 new boxes to $\mathcal{L}$, $n = n + 2$, $i = i + 1$, goto 3

All the steps of this algorithm have already been described in the previous section except for the choice of the bisected variable. The proposed strategy is to bisect the variable having the largest interval except if all the derivatives of the polynomial with respect to the parameters are available, in which case Kearfott's maximal smear function is used [8].

With a minimal extension of the previous algorithm the following problems may be solved:

- calculation of both the minimal and maximal roots
- calculation of the minimal and maximal condition number of $J^T J$ (ratio smallest eigenvalue/ largest eigenvalue)
- additional constraints on the parameters can be taken into account (see example)

7   Largest Parameters Region with Real Part of the Roots in a Given Range

The problem here is to find an approximation of the region(s) $\mathcal{R}$ in the $X = \{x_1, x_2, \ldots, x_n\}$ space such that all the polynomials represented by a point in this region have all their roots with a real part in a given range $[a, b]$.

The structure of the algorithm will be similar to the one mentioned in the previous section but with a modification on the maximal root value test.

7.1   Root Bounds Test

The purpose of this test will be to check if a given box may include parameter values such that the corresponding polynomial has a root with a real part larger than b or lower than a. It may return *true* (all the polynomials have at least one root with real part lower than a or larger than b), *false* (all the polynomial have their roots with a real part in $[a, b]$) or *indeterminate* if none of the 2 above cases applies.

This test uses various heuristics: some of them, denoted `Real`, are used only if we are investigating the real roots while others, denoted `RealPart`, are used only if we are considering the real part of the root. Heuristics that may be used in both cases will have no specific notation:

- (`Real`)interval evaluation of $P(] - \infty, a[)$, $P(]b, \infty[)$: if both evaluations do not include 0 the algorithm will return *false*,
- (`RealPart`) interval evaluation of the real and imaginary parts of $P(] - \infty, a[+IC)$, $P(]b, \infty[+IC)$ where C has lower bound 0 and upper bound `MaxRoot`(P): if one of the evaluations does not include 0 the algorithm will return *false*,
- the Budan-Fourier method for the ranges $] - \infty, a]$, $[b, \infty[$: if this method returns $[0,0]$ for both ranges the test will return *false* (`Real`). If the method returns an interval of type $[b, b]$ where b is an odd number for at least one of the ranges, then the test will return *true*
- Kharitonov polynomials: this method is used to determine if all the roots of P are greater or equal to a and lower or equal to b in which case the test will return *false*. Alternatively the test will return *true* if the Kharitonov polynomials allows to show that all the roots are larger than b or lower than a
- Routh table: all the polynomial with roots having a real part larger than a must be such that the polynomial in y $P^{\star}(y) = P(y + a)$ has a Routh table whose first column elements exhibit a number of sign changes equal to the degree of the polynomial. Alternatively all the polynomials with roots having a real part lower than b must be such that the polynomial in y $P^{\star}(y) = P(y + b)$ has a Routh table whose first column elements have the same sign. In the both cases we may use the 2B method to filter the box. Assume for example that element R_{j1} at the j-th row of the first column of the Routh table of the polynomial $P(y + b)$ has a constant sign (e.g. is positive). The parameters belonging to the valid region must be such that $R_{k1} > 0$ for all k. Assume now for example that R_{k1} may be written as $x_1 + F(\mathbf{X})$ where x_1 is one of the parameters. The relation $x_1 + F(\mathbf{X}) > 0$ implies $x_1 > -F(\mathbf{X})$ and let U_2 be the interval evaluation of the right side of this inequality. This inequality imply that $\underline{x_1} > \overline{U_2}$ that may allow to update the value of $\underline{x_1}$.

8   Algorithm

We will now explain an algorithm whose purpose is to determine an approximation R of $\mathcal{R}$ together with a measure of the quality of the approximation.

The structure of the algorithm is classical with a list $\mathcal{L}$ of n boxes B_j, indexed by i. The output of the algorithm is a solution list of boxes such that for any parameter values inside the boxes the real roots of the corresponding polynomial lie in the range $[a, b]$. The total volume of neglected boxes will be denoted V_n.

1. initialization of the range for $z=[\underline{Z}, \overline{Z}]$,
2. creation of the first box, $i = 1$, $\mathcal{L} = \{B_1\}$, $n = 1$
3. if $i > n$, then exit
4. if check $P(B_i)$=*false*, then store B_i in the solution list, $i = i + 1$, goto 3
5. if check $P(B_i)$=*true*, then $i = i + 1$, goto 3
6. if the width of B_i is lower than a threshold ϵ reject box and add its volume to V_n
7. bisect B_i, add the new boxes to $\mathcal{L}$, $n = n + 2$, $i = i + 1$, goto 3

The solution list is only an approximation of $\mathcal{R}$ as boxes with a width lower than ϵ for which the root bounds test has not returned either *true* or *false* are not processed.

Note that the quality of the approximation may be measured by the ratio (volume of R)/V_n denoted as the *quality index*. Furthermore an interesting point is that the approximation R can be computed *incrementally*. Indeed we may start the algorithm with a large value of ϵ and store the boxes that are rejected due to their size in a file. Then, if the quality index is not satisfactory, we may restart the algorithm with a lower ϵ and $\mathcal{L}$ being initialized with the set of boxes that have been rejected during the previous run of the algorithm (thereby avoiding to repeat a large number of calculations).

Note also that a similar algorithm has been designed to determine the largest hypercube in the parameters space such that all the real roots of the polynomial having a representative point in this hypercube lie in the range $[a, b]$.

9 Implementation and Maple Interface

9.1 ALIAS

All of the previously described algorithms are implemented in the C++ ALIAS library developed in the COPRIN project[1]. ALIAS is a collection of interval analysis based methods for system solving, optimization, ... which is freely available for SUN/PC platforms running Unix and Linux. In this package we use a corrected version of the Bias/Profil library for the basic interval operations (although on the long term we plan to switch to the multi-precision interval package MPFI that will allow for faster and more accurate interval calculation).

Although the C++ ALIAS library may be directly used to develop algorithms we offer another interface through Maple. Within a Maple session the end-user may define a polynomial or the matrix and then call specific Maple procedures of the ALIAS Maple interface. These procedures will create the necessary C++ code according to the problem to be solved, compile it and then run the executable. After the completion of the calculations the results are written in files and are made available directly inside the same Maple session. Note also that for algorithms in the library there is a distributed implementation that allows to run the algorithm on a set of computers without ever leaving the Maple session.

---

[1] www-sop.inria.fr/coprin/logiciels/ALIAS/ALIAS.html

10   Symbolic Computation and Efficiency

It may be noted that there is nothing original in the structure of the algorithm. As for most interval analysis based algorithm the efficiency is drastically influenced by the quality of the heuristics that are used during the calculation. A strong point of our implementation is adaptation of most classical algebraic geometry theorems to interval analysis: bounds for the roots (over 15 different operators are used to determine these bounds), Budan-Fourier method, ...

But the purpose of this section is to explain why symbolic computation allows a drastic improvement of the efficiency.

10.1   Complex Parametric Matrices

When dealing with parametric matrices a first problem is to determine an analytic form for the coefficients of the characteristic polynomial which is needed for most of our heuristics. Indeed as soon as the dimension of the matrix exceed a relatively small number (typically 6) it is difficult, even with symbolic computation, to obtain these coefficients. Interval evaluation of these coefficients may be obtained numerically from the analytical expressions of the components of the matrix but such calculation is usually inefficient and leads to very large overestimation of the coefficients. but this is not completely true due to the use of Maple as an interface.

Indeed consider a matrix A of dimension n and assume, for example, that a sub-matrix of the matrix $A - zI_n$ (where I_n is the identity matrix) is:

$$\begin{pmatrix} 3x_1 - z & x_1 \\ 6 & 2 - z \end{pmatrix}$$

During the calculation of the coefficients of P we will use the determinant of this sub-matrix. Numerically this determinant will be computed as $z^2 - z(3x_1 - 2) + 6x_1 - 6x_1$. Hence note that the constant term of this polynomial in z does not evaluate to 0. Our Maple procedure allows to compute symbolically all the minors of the matrix A up to a given dimension and to use a C++ interval evaluation of the coefficients of these polynomials in the calculation of the coefficients of P. In our example the determinant of the minor will be evaluated as $z^2 - z(3x_1 - 2)$ and this correct expression will be used for the computation of the coefficients of P.

10.2   Improving the Interval Evaluation of Expressions

Maple is not only used to create C++ code. First of all there is a semantic analysis of the expressions in order to create an efficient evaluation form for the expressions (factorization, compactness . . .). Indeed it is well known that the analytical form that is used for interval evaluation has a drastic influence on the over-estimation of the interval. We have developed a specific Maple procedure which tries to determine what is, on the average, the best analytical form of an expression : basically this procedure tries to factor the expression and convert it into Horner's form, using different orders for the unknown, until a form whose evaluation involves the lowest number of operations or which has only one occurrence of the unknown (in which case the interval evaluation is exact [10]) is found. Although theoretical work has to be done for this procedure it already gives interesting results, especially for large expressions that anyway cannot be dealt with by hand.

In a similar manner there is an optional use of the derivative of the expression that allows to improve the interval evaluation either by using its monotonicity or by calculating the interval evaluation using its Taylor form.

10.3   Kharitonov Polynomials and Routh Table

As mentioned in the previous section we need to determine the coefficients of the polynomial in y $P^*(y) = P(y + b)$ where b is the lower or upper bound of the allowed range for the real part of the roots of P. These coefficients may be derived numerically from the coefficients a_i but a Maple expansion beforehand may allow for possible simplifications and better interval evaluation of these coefficients.

Consider for example the polynomial

$$P = x_1 z^2 + 2x_1 z + 1$$

The coefficient of y in the P^* polynomial will be computed numerically as the interval resulting from the evaluation of $2x_1 - 2ax_1$, while this coefficient will be calculated as $2x_1(1 - a)$ with a Maple expansion. The first case may lead to an over-estimation of the interval as the variable x_1 appears twice in the expression, while in the second case x_1 appears only once and hence we get the exact bounds for this coefficient.

11   Application Examples

11.1   Robotics

We have considered the Orthoglide [13] parallel robot developed at IRCYN (Nantes) which is a 3 d.o.f. cartesian robot (the end-effector of the robot moves along the x, y, z axis while there is no change in the orientation of the end-effector).

The characteristic polynomial of the $J^T J$ matrix is of degree 3. Besides the accuracy analysis the interest in the eigenvalues of this matrix is that they allow to determine how efficiently the motors are used. Indeed they represent the amplification factor between each motor and the forces exerted by the platform. If an eigenvalue is close to 0 the power of the motor is used mostly to preserve the geometry of the structure and does not contribute to the force provided at the end-effector level.

The characteristic polynomial has 101 terms and its coefficients are a mix of algebraic terms and square root terms. More specifically the square roots of the following terms are involved: $1 - x^2 - y^2$, $1 - x^2 - z^2$, $1 - y^2 - z^2$. Consequently we have 3 additional constraints:

$$x^2 + y^2 \leq 1 \quad , \quad x^2 + z^2 \leq 1 \quad , \quad z^2 + y^2 \leq 1 \quad , \tag{4}$$

which describe that the workspace of the robot is the intersection of three cylinders that are directed along the x, y, z axis.

For this robot the following problems have to be solved:

– determination of the lowest and largest eigenvalues over the whole workspace of the robot,
– calculation of the largest cube enclosed in the workspace such that for any pose in the cube the eigenvalues are in the range $[0.25, 4]$,
– calculation of the regions in the workspace such that for any pose in the region the eigenvalues are in the range $[0.25, 4]$,

All these calculations have been done using the **ALIAS** Maple interface. A typical call to this library is:

```
MinMax_Polynom_Area([eq1,eq2,eq3,eq],[lambda,x,y,z],
                    [[-1000,1000],[-1,1],[-1,1],[-1,1]],
                    [0,3000],3,30,10,"Out",0.05);
```

where `eq1,eq2,eq3` are the constraints described in equations (4) and `eq` is the characteristic polynomial. The unknown in the polynomial is `lambda` and `x`, `y`, `z` are the parameters. The next argument is a set of ranges for each unknown: the first range is the range for the polynomial unknown and is automatically adjusted during the calculation. The following arguments are parameters for the method: for example the string `Out` give the name of the file in which the boxes of the solution list are written while the last argument is the threshold ϵ on the width of the boxes that will be rejected.

The following computation times have been obtained on a Sun Blade workstation:

– minimal and maximal roots: less than 3 s with an accuracy 10^{-3}
– largest cube where the roots are in $[0.25, 4]$: 1 minute
– region where the roots are in $[0.25, 4]$ with $\epsilon = 0.05$: 30487 seconds. The volume of the approximated region is 1.32533 while the total volume of the neglected boxes is 1.488. Figure 2 shows a cross-section at $z = 0.5$ of the approximation R (the gray boxes) and the boxes that have been rejected (black boxes). It can be seen on the figure that a first class of rejected boxes corresponds either to the boundary of the workspace or to the boundary of the region R while a second class indicates a possible extension of the approximation R.

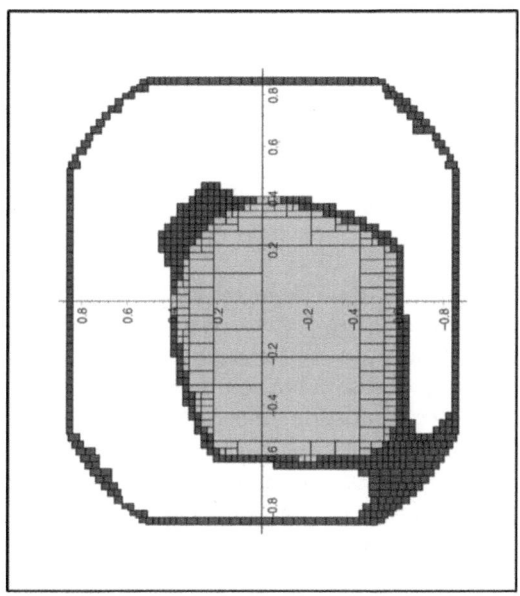

Fig. 2. Cross-section in the $x - y$ plane of the region where all the polynomials have their roots in the range [0.25,4] computed with $\epsilon = 0.05$ (in gray R and in black the rejected boxes). The area of R is 0.847 while the area of the rejected boxes is 0.3987

11.2   Control Theory: Robust Mobile Robot Path Planning

Another problem that has been considered is the robust control of a mobile robot along a nominal trajectory which is not exactly followed by the robot due to uncertainties (slipping of the wheels, friction). Robust control is achieved if the minimal eigenvalue of a controllability matrix is positive whatever is the location of the robot in a tube around its nominal trajectory.

The controllability matrix is algebraic:

$$\begin{bmatrix} 20\,f & 0.1\,s & 10\,fx + 0.1\,sy + 0.1\,u - 25 - x \\ 0.1\,s & 0.2\,fr & -0.25 + 0.1\,fry + 0.001\,v - y \\ 10\,fx + 0.1\,sy + \\ 0.1\,u - 25 - x & -0.25 + 0.1\,fry + \\ 0.001\,v - y & 2 \end{bmatrix}$$

with the following constraints:

$$\left(x + 1.05 - \cos(\frac{\pi\,t}{10})\right)^2 \le 0.0001$$

$$\left(y + 0.05 - \sin(\frac{\pi\,t}{10})\right)^2 \le 0.0001$$

$$\left(z - 0.1 + 0.1\ \cos(\frac{\pi\,t}{10})\right)^2 \le 0.0001$$

where v, r, f, x, y, z, s, u are parameters and t is the time supposed to lie in the range $[0,10]$. The ranges for the parameters are:

$$v \in [200, 300] \qquad r \in [0.8, 1.2] \qquad f \in [800, 1200]$$
$$s \in [-20, 20] \qquad u \in [200, 300]$$

while the range for x, y, z may be deduced from the constraint equations. With these ranges the system is badly conditioned as there are variables that have a very low influence on the value of the eigenvalues. Hence the use of the smear function is essential to get reasonable computation time.

With $\epsilon = 10^{-4}$ the computation time is about 1 hour on a 800 Mhz PC to determine that the system is controllable as the lowest and largest eigenvalue are 1.03573 and 3.146889; this time is reduced to less than 10 minutes if ten computers are used. Note that here we may have considered computing only the lowest eigenvalue but the largest one gives also useful information on the system.

11.3   Control Theory: Helicopter

This problem is classical in the field of robust control [5]. Let a, x, y be three polynomials in the variables s:

$$a = s\ (0.1\ s + 1)\ (s + Q2)$$
$$y = y_0 + y_1\ s + y_2\ s^2$$
$$x = x_0 + x_1\ s + s^2$$

and consider the polynomial P defined as

$$P = a\ x + Q1\ y \tag{5}$$

In this polynomial the parameters are x_0, x_1, y_0, y_1, y_2, while $Q1, Q2$ are physical parameters that may lie in the range $[20,60]$, $[0,1]$.

To have a robust control of the helicopter it is necessary to ensure that the real parts of the roots of P are all negative. In the classical approach using convex optimization the problem is solved only for the 4 corners of the $Q1, Q2$ box. We have investigated the determination of all the possible values of the parameters ensuring a robust control over the whole $Q1, Q2$ box. Using a parallel implementation of ALIAS which relies on 12 computers we have obtained the result presented in table (1) which gives the width ϵ of the neglected box, the total volume of the solution boxes, the total volume of the neglected boxes and their number n. The initial range for all the variables was $[0.2,10]$ (hence a search space of volume 90392.07968).

The total computation time up to accuracy 0.3125 was 40 minutes. As may be seen solutions are obtained rather early in the computation. But the quality

Table 1. Solution volume as function of the maximal width of the neglected boxes

ϵ	Solution	Neglected	n
10	0	30438	1
5	0	9887	8
2.5	0	6433	92
1.25	0	4647	1147
0.625	0.517506	2751	18248
0.3125	25.648522	1610	331916

of the approximation at the end of the calculation is not satisfactory. It will be possible to improve it but the number of neglected boxes at the last step is already large.

12   Conclusion

Dealing with polynomials having uncertain coefficients is very important for a large number of applications such as robotics and control theory, ...

Algorithms based on interval analysis are an interesting alternative to the more classical approach. They offer richer results such as finding not only a solution but an approximation of the set of all solutions or dealing with uncertainties in the polynomial and have proven to be effective on realistic examples.

The main weakness of the proposed algorithms is probably that we have not sufficiently investigated the potentialities of matrix computations (preconditioning, decomposition, ...) that may improve their efficiency.

References

[1] Ackermann J. and others . *Robust control: systems with uncertain physical parameters*. Springer-Verlag, 1994.

[2] Ciarlet P. and Lions J-L. *Handbook of numerical analysis, 7 : solution of equations in Rn (part 3)*. North-Holland, 2000.

[3] Collavizza M., F. Deloble, and Rueher M. Comparing partial consistencies. *Reliable Computing*, 5:1–16, 1999.

[4] Golub G. H. and Van Loan C. F. *Matrix computations*. John Hopkins University Press, 1996.

[5] Henrion D. Stabilisation robuste de systèmes polytopiques par contrôleur d'ordre fixe avec la boîte à outils polynomiale pour matlab. In *Conférence Internationale Francophone d'Automatique*, pages 682–687, Nantes, June, 8-10, 2001.

[6] Isaacson E. and Keller H. B. *Analysis of numerical methods*. John Wiley, 1966.

[7] Jaulin L., Kieffer M., Didrit O., and Walter E. *Applied Interval Analysis*. Springer-Verlag, 2001.

[8] Kearfott R. B. and Manuel N. III. INTBIS, a portable interval Newton/Bisection package. *ACM Trans. on Mathematical Software*, 16(2):152–157, June 1990.

[9] Mignotte M. *Mathématiques pour le calcul formel.* PUF, 1989.

[10] Moore R. E. *Methods and Applications of Interval Analysis.* SIAM Studies in Applied Mathematics, 1979.

[11] V. L. Kharitonov. Asymptotic stability of an equilibrium position of a family of systems of linear differential equations. *Differential'ye Uravneniya*, 14:2086–2088, 1978.

[12] Weispfenning V. Solving parametric polynomial equations and inequalities by symbolic algorithms. In *Computer Algebra in Science and Engineering*, pages 163–179, Bielefeld, August 1994. World Scientific.

[13] Wenger P. and Chablat D. Kinematic analysis of a new parallel machine-tool: the Orthoglide. In *ARK*, pages 305–314, Piran, June, 25-29, 2000.

[14] Zippel R. *Effective polynomial computation.* Kluwer, 1993.

D.C. Programming for Solving a Class of Global Optimization Problems via Reformulation by Exact Penalty

Le Thi Hoai An

Laboratory of Modelling, Optimization & Operations Research
National Institute for Applied Sciences-Rouen
BP 8, F 76 131 Mont Saint Aignan Cedex, France
`lethi@insa-rouen.fr`

Abstract. We consider a class of important problems in global optimization, namely concave minimization over a bounded polyhedral convex set with an additional reverse convex constraint. Using a related exact penalty property, we reformulate the class of mixed zero-one concave minimization programs, concave minimization programs over efficient sets, bilevel programs, and concave minimization programs with mixed linear complementarity constraints in the form of equivalent d.c. (difference of convex functions) programs. Solution methods based on d.c. optimization algorithms (DCA) and the combined DCA - branch and bound are investigated.

Keywords: d.c. programming, polyhedral d.c. programming, DCA, combined DCA - branch and bound algorithm, reformulations, exact penalty

1 Introduction

Let K be a nonempty bounded polyhedral convex set in $\mathbb{R}^n$ and f, p be concave functions on K. We will be concerned with the nonconvex program

$$\alpha = \inf\{f(x) : x \in K, p(x) \leq 0\}, \tag{1}$$

where the nonconvexity appears both in the objective function and the constraint $\{x \in \mathbb{R}^n : -p(x) \geq 0\}$ called the reverse convex constraint. To make solution methods for (P) efficient we try to penalize the reverse convex constraint. For example, the usual exact penalty process leads to

$$\alpha_t = \inf\{f(x) + tp^+(x) : x \in K\}, \tag{2}$$

where t is a positive number (called penalty parameter) and $p^+(x) = \max(0, p(x))$. Actually the exact penalty holds if there is a positive number t_o such that for $t \geq t_o$ the solution sets of (1) and (2) are identical. Such a property is generally true in convex optimization. However, for nonconvex optimization it becomes more complex and therefore a specific study should be carried out.

C. Bliek et al. (Eds.): COCOS 2002, LNCS 2861, pp. 87–101, 2003.

The exact penalty technique aims at transforming (1) into a more tractable equivalent problem of the d.c. optimization framework. A d.c. program is of the form

$$\inf\{f(x) = g(x) - h(x) : x \in \mathbb{R}^n\} \qquad (3)$$

with g, $h \in \Gamma_o(\mathbb{R}^n)$, the set of all lower semi-continuous proper convex functions on $\mathbb{R}^n$. Such a function f is called d.c. function, and g, h are its d.c. components. D.c. programming plays a key role in nonconvex optimization and covers almost real world nonconvex programs. It has been extensively developed in recent years from both combinatorial approaches [10], [11], [23], [24], [25], [26], and convex analysis approaches [1] - [6], [15], [16].

It has been proved in [5] that the exact penalty property for (1) holds in the case where the function p is nonnegative over K. This result allows to re-formulate a class of important and difficult nonconvex programs in the form of d.c. programs. They are mixed zero-one concave minimization programs, con-cave minimization programs over efficient sets, bilevel programs, and the concave minimization programs with linear complementarity constraints. In this paper we develop d.c. optimization approaches based on decomposition branch and bound techniques and DCA for solving these problems.

2   D.C. Reformulation via the Exact Penalty

Our reformulations are based on the following result:

Theorem 1 *Let K be a nonempty bounded polyhedral convex set in $\mathbb{R}^n$ and f, p be finite concave on K. Assume the feasible set of (P) be nonempty and p be nonnegative on K. Then there exists $t_o \geq 0$ such that for every $t > t_o$ the following problems have the same solution sets:*

$$(P_t) \quad \alpha_t = \inf\{f(x) + tp(x) : x \in K\},$$

$$(P) \quad \alpha = \inf\{f(x) : x \in K, p(x) \leq 0\}.$$

Furthermore
(i) If $p(x) \leq 0$ for any x in the vertex set of K, denoted by $V(K)$, then $t_o = 0$.
(ii) If $p(x) > 0$ for some x in $V(K)$, then $t_o \leq \frac{f(x^o) - \alpha(0)}{S}$ for every $x^o \in K$ stisfying $p(x^o) \leq 0$, where $S := \min\{p(x) : x \in V(K), p(x) > 0\}$.

Proof. See [5]

Remark 1 *(i) If p is concave on K and if the feasible set of (P) is nonempty, then $V(K) \cap \{x \in K : p(x) \leq 0\} \neq \emptyset$.*
(ii) If f is concave on K and $V(K) \subset \{x \in K : p(x) \leq 0\}$, then $\alpha(0) = \alpha$. In this case the constraint $p(x) \leq 0$ is not essential for (P).
(iii) Under the assumptions of Theorem 1, we have

$$\alpha = \min\{f(x) : x \in K, p(x) \leq 0\} = \min\{f(x) : x \in V(K), p(x) \leq 0\}.$$

It is worth noting that the exact penalty approach for (P) is nothing but the Lagrangian duality relative to this problem. Indeed, the Lagrangian duality relative to (P) introduces Problem (P$_t$), and the dual problem of (P) is given by

$$\text{(D)} \quad \beta = \sup\{\alpha(t) : t \in \mathbb{R}^+\}.$$

From Theorem 1 it follows that there is no duality gap, and the solution set of (P) can be deduced from the solution of its dual program (D) as in convex optimization. More precisely, the solution set of (D) is actually the set of exact penalty parameters, say $]t_o, +\infty[$.

Theorem 1 permits to transfer (P) into a d.c. program. Indeed, if the functions f and p are finite concave on a nonempty bounded polyhedral convex set K and if p is nonnegative on K, then (P$_t$) and (2) are identical and are d.c. programs:

$$\alpha(t) = \min\{f(x) + tp(x) : x \in K\} = \min\{\chi_K(x) - [-f(x) - tp(x)] : x \in \mathbb{R}^n\}$$
$$= \min\{g(x) - h(x) : x \in \mathbb{R}^n\} \text{ with } h = -(f + tp) \text{ and } g = \chi_K.$$

Here χ_K stands for the indicator function of K: $\chi_K(x) = 0$:if $x \in K$, $+\infty$ otherwise.

3 Numerical Solution Methods via D.C. Programming

In recent years, d.c. programming has been developed extensively, becoming an attractive topic of research in nonconvex programming. In this section we study two approaches in both global and local approaches to solving (P$_t$): a d.c. optimization algorithm called DCA and a branch-and-bound scheme.

3.1 D.C. Optimization Algorithm (DCA)

First we summarize the background material needed for presenting d.c. programming and DCA. We follow [18] for definitions of usual tools of convex analysis where functions could take the infinite values $\pm\infty$. A function $\theta : X \to \mathbb{R} \cup \{\pm\infty\}$ is said to be proper if it takes nowhere the value $-\infty$ and is not identically equal to $+\infty$. For $g \in \Gamma_0(X)$, the *conjugate function* g^* of g is a function belonging to $\Gamma_0(Y)$ and defined by

$$g^*(y) = \sup\{\langle x, y \rangle - g(x) : x \in X\}.$$

For $g \in \Gamma_0(X)$ we have $g^{**} = g$.

Let $g \in \Gamma_0(X)$ and let $x^0 \in \text{dom } g$ and $\epsilon > 0$. Then $\partial_\epsilon g(x^0)$ stands for the $\epsilon-$ *subdifferential* of g at x^0 and is given by

$$\partial_\epsilon g(x^0) = \{y^0 \in Y : g(x) \geq g(x^0) + \langle x - x^0, y^0 \rangle - \epsilon, \forall x \in X\}$$

while $\partial g(x^0)$ corresponding to $\epsilon = 0$ stands for the *usual (or exact) subdifferential* of g at x^0.

A point x^* is said to be *a local minimizer* of $g - h$ if $g(x^*) - h(x^*)$ is finite (i.e., $x^* \in \text{dom } g \cap \text{dom } h$) and there exists a neighbourhood U of x^* such that

$$g(x^*) - h(x^*) \leq g(x) - h(x), \quad \forall x \in U. \tag{4}$$

Under the convention $+\infty - (+\infty) = +\infty$, the property (4) is equivalent to $g(x^*) - h(x^*) \leq g(x) - h(x), \quad \forall x \in U \cap \text{dom } g$.

x^* is said to be *a critical point* of $g - h$ if $\partial g(x^*) \cap \partial h(x^*) \neq \emptyset$.

We shall work with the space $X = \mathbb{R}^n$ which is equipped with the canonical inner product $\langle \cdot, \cdot \rangle$ and the corresponding Euclidean norm $\| \cdot \|$, thus the dual space Y of X can be identified with X itself. The DCA was introduced by Pham Dinh in 1986 [13, 14] as an extension of his subgradient algorithms (for convex maximization programming) to d.c. programming. However, this field has been really developed since 1994 by the joint work of Le Thi and Pham Dinh (see, e.g. Le Thi [1], Le Thi - Pham Dinh [2] - [6],[15, 16] and references therein). The DCA is a primal-dual subdifferential method for solving a general d.c. program of the form

$$(P_{dc}) \qquad \alpha = \inf\{f(x) := g(x) - h(x) : x \in \mathbb{R}^n\}$$

with $g, h \in \Gamma_o(\mathbb{R}^n)$ and its dual program

$$(D_{dc}) \qquad \alpha = \inf\{h^*(y) - g^*(y) : y \in \mathbb{R}^n\}.$$

Ones say that $g - h$ is a d.c. decomposition of f, and g, h are its d.c. components. We note that any convex constrained d.c. program can be written in the standard form (P_{dc}) by adding the indicator function of the convex set of constraints in the objective function.

A function $\theta \in \Gamma_o(X)$ is said to be polyhedral convex if ([18])

$$\theta(x) = \max\{\langle a_i, x \rangle - \alpha_i\} : i = 1, \ldots, m\} + \chi_K(x) \quad \forall x \in X,$$

where K is a nonempty polyhedral convex set in X. If either g or h is a polyhedral convex function, then ones say that (P_{dc}) is a polyhedral d.c. program. The class of polyhedral d.c. programs, which plays a key role in nonconvex optimization, possesses worthy properties from both theoretical and computational viewpoints, as necessary and sufficient local optimality conditions, and finite convergence for DCA.

Based on the d.c. duality and the local optimality, the DCA consists of the construction of two sequences $\{x^k\}$ and $\{y^k\}$ such that x^{k+1} (resp. y^k) is a solution to the convex program (P_k) (resp. (D_k)), where (P_k) (resp. (D_k)) is obtained from (P_{dc}) (resp. (D_{dc})) by replacing h (resp. g^*) with its affine minorization:

$$(P_k) \qquad \{\inf\{g(x) - [h(x^k) + \langle x - x^k, y^k \rangle] : x \in \mathbb{R}^n\},$$

$$(D_k) \qquad \{\inf\{h^*(y) - [g^*(y^{k-1}) + \langle x^k, y - y^{k-1} \rangle] : y \in \mathbb{R}^n\}.$$

In other words, the DCA yields the scheme:

$$y^k \in \partial h(x^k); \quad x^{k+1} \in \partial g^*(y^k). \tag{5}$$

By this way, the sequences $\{g(x^k) - h(x^k)\}$ and $\{h^*(y^k) - g^*(y^k)\}$ are decreasing in an appropriate way and the corresponding limits x^∞ and y^∞ of $\{x^k\}$ and $\{y^k\}$ satisfy the local optimality condition

$$\partial h(x^\infty) \subset \partial g(x^\infty) \quad and \quad \partial g^*(y^\infty) \subset \partial h^*(y^\infty),$$

or they are critical points of $g - h$ and $h^* - g^*$, respectively (see [1, 2, 15, 16]).

The crucial feature of the DCA is the *choice* of a *good* d.c. decomposition and a *good* initial point that are open questions to be studied. Of course, this depends strongly on the very specific structure of the problem being considered. In practice, for solving a given d.c. program, we try to choose g and h such that sequences $\{x^k\}$ and $\{y^k\}$ can be easily calculated, i.e. either they are in explicit form or their computations are inexpensive. For a detailed study of d.c. programming and DCA we refer the readers to [1, 15, 16].

Let us consider now Problem (P_t):

$$(P_t) \quad \alpha_t = \inf_{x \in K} \{f(x) + tp(x)\} = \inf_{x \in \mathbb{R}^n} \{F_t(x) := \chi_K(x) + f(x) + tp(x)\}.$$

Assume that subgradients of $-f$ and of $-p$ are computable, a d.c. decomposition of F_t can be chosen as shown in Section 2, namely

$$F_t(x) := g(x) - h(x) \text{ with } g(x) := \chi_K(x), h(x) := -f(x) - tp(x). \qquad (6)$$

In this case (P_t) is a polyhedral d.c. program because χ_K is a polyhedral convex function, and the general DCA scheme (5) becomes:

$$y^k \in \partial(-f(x^k) - tp(x^k)); \quad x^{k+1} \in \arg\min\{-\langle x, y^k \rangle : x \in K\}. \qquad (7)$$

Besides the computation of subgradients of $-f$ and $-p$, the algorithm requires one linear program at each iteration and it has a finite convergence. The convergence properties can be stated as follows:

Theorem 2 *i)If x^k is chosen as a vertex in (7) DCA generates a finite sequence $x^1, ..., x^{k_*}$ contained in $V(K)$ such that $f(x^{k+1}) + tp(x^{k+1}) \leq f(x^k) + tp(x^k)$ for each k, and x^{k_*} is a critical point of $g - h$.*
ii) If, in addition, h is differentiable at x^{k_*}, then x^{k_*} is actually a local minimizer to (P_t).
iii) Let $t > t_1 := \max\left\{\frac{f(x)-\alpha(0)}{S} : x \in V(K) \cap \{x \in K : p(x) = 0\}\right\}$. If at an iteration l one has $p(x^l) = 0$, then $p(x^k) = 0$ and $f(x^{k+1}) \leq f(x^k)$ for all $k \geq l$.

Proof. Properties i) and ii) are direct consequences of DCA's Convergence Theorems for general polyhedral d.c. programs, see Theorems 3, 5, 6 of [16]. Only iii) needs a proof. If $V(K) \subset K_p := \{x \in K : p(x) = 0\}$, then the assertion is trivial. Assume now that $V(K)$ is not contained in K_p. From i) of this theorem it follows that

$$f(x^{k+1}) + tp(x^{k+1}) \leq f(x^k) + tp(x^k). \qquad (8)$$

If $x^l \in K_p$, i.e., $p(x^l) = 0$, then (8) implies that

$$tp(x^{l+1}) \le f(x^l) - f(x^{l+1}). \tag{9}$$

As a consequence, if $p(x^{l+1}) > 0$, then

$$t \le \frac{f(x^l) - f(x^{l+1})}{p(x^{l+1})} \le \frac{f(x^l) - \alpha(0)}{S} \le t_1.$$

This means that for any $t > t_1$ we have $p(x^{l+1}) = 0$, i.e., $x^{l+1} \in K_p$. On the other hand, from (9) we have $f(x^l) - f(x^{l+1}) \ge 0$. The proof is then complete.

Remark 2 *Since DCA is a descent method, it enjoys the interesting property iii) which is very important when using the exact penalty approach for solving (P): in practice, global optimization methods such as branch and bound applied to (P_t) are often used to find an ε-solution x_ε, since it is very difficult to obtain an exact solution. Unfortunately, such an ε-solution of (P_t) may be not feasible to (P) (when $p(x_\varepsilon) > 0$). However, due to Property iii), the DCA applied to (P_t) with t sufficiently large provides a feasible solution to (P). This advantage of the DCA makes our combined DCA-branch and bound method efficient in finding an ε-solution of (P) in the large scale setting. See, for example, Remark 3 concerning the quadratic zero-one programming.*

3.2   The Combined DCA - Branch and Bound Method

Theorem 2 proves that in many cases DCA provides a local minimizer to (P_t). Then is it very interesting to introduce DCA in branch and bound algorithms for obtaining a global minimizer. Using DCA for computing upper bounds, our combined DCA branch-and-bound algorithm to solve d.c. programs

$$\eta := \min \{ f(x) := g(x) - h(x) : x \in C \} \tag{10}$$

(where f is a real valued continuous function and C is a compact set) can be stated as follows: ($\mathcal{R}$ is the collection of the subsets that to be considered in the algorithm)

Initialization:
Let $M \supset C$. Set $\{M\} \leftarrow \mathcal{R}$. Compute a lower bound $\beta(M)$ for η.
Solve (10) by DCA to obtain a point x^1, set $\gamma := f(x^1)$.
Let $k \leftarrow 1$, $\beta \longleftarrow \beta(M)$
Iteration k=1, 2, ...
while $\gamma > \beta$ (resp. $\gamma > \beta + \varepsilon$) **do**
k1. Choose $M_k \in \mathcal{R}$ such that $\beta(M_k) = \beta$.
k2. Divide M_k into r subsets $M_k^1, \ldots, M_k^r$ satisfying

$$\bigcup_{i=1}^{r} M_k^i = M_k, \quad \text{int } M_k^i \cap \text{ int } M_k^j = \emptyset \quad \text{for } i \ne j;$$

For each $i = 1, \ldots, r$ compute a lower bound $\beta(M_k^i)$ of f on M_k^i
k3. Replace $\mathcal{R}$ by $\mathcal{R} = \mathcal{R} \cup_{i=1}^r M_k^i \setminus (\{M_k^i : \beta(M_k^i) \geq \gamma\} \cup M_k)$.
(resp. $\mathcal{R} = \mathcal{R} \cup_{i=1}^r M_k^i \setminus (\{M_k^i : \beta(M_k^i) \geq \gamma - \varepsilon\} \cup M_k)$.
Set $\beta \longleftarrow \min\{\beta(M) : M \in \mathcal{R}\}$,
k4. If a point $x' \in C$ such that $f(x') < f(x^k)$ is detected, then
update the upper bound by applying DCA from x' to obtain $\overline{x}$, set
$\gamma \longleftarrow f(\overline{x})$, $x^k \longleftarrow \overline{x}$.
$k \leftarrow k + 1$
endwhile
   x^k is optimal solution (resp. ε-solution) to Problem (10).

The use of the branch and bound method depends on the structure of the problem under consideration. For Problem (10) we often use the convex envelope θ of the concave function $-h$ while computing lower bounds. More precisely, a lower bound $\beta(M_i)$ of $g - h$ on M_i is

$$\beta(M_i) = \min \{g(x) + \theta(x) : x \in M_i\}. \tag{11}$$

We have to choose d.c. components g and h such that θ can be computed easily and the convex subproblems (11) can be solved efficiently. For example, if f is twice continuously differentiable, then we choose h as follows:

$$f(x) = g(x) - h(x) := \left(\frac{\rho}{2}\|x\|^2 + f(x)\right) - \frac{\rho}{2}\|x\|^2,$$

where ρ is positive number such that the function $\frac{\rho}{2}\|.\|^2 + f$ is convex. Then the convex envelope θ of $-\frac{\rho}{2}\|.\|^2$ in the box $M_i := \overline{\Pi}_1^n [a_j, b_j]$ s simply the sum of affine functions $l_j(x_j)$ that agree with $(-1/2)\rho x_j^2$ at the ends of the segment $[a_j, b_j]$. Namely,

$$\theta(x) = \sum_{j=1}^n l_j(x_j) = -\frac{\rho}{2}\sum_{j=1}^n [(a_j + b_j)x_j - a_j b_j].$$

For the branching procedure the most useful is the rectangular or simplicial subdivision (see e.g. [11]).

4   Applications

We study in this section a class of nonconvex problems that can be formulated as (P) and then are equivalent to d.c. programs of the form (P_t), according to Theorem 1. We show how to apply the general DCA scheme (7) to these problems for sufficiently large values of t. A detailed discussion about the branch and bound algorithms will be presented in a future work.

4.1   Mixed Zero-One Concave Minimization Programs

Let D be a nonempty bounded polyhedral convex set in $\mathbb{R}^n$ and let I be a given subset of $\{1, \ldots, n\}$. The mixed zero-one concave minimization programming problem is defined as

$$\min \{f(x) : x \in D,\ x_i \in \{0,1\}, \forall i \in I\}, \tag{12}$$

where f is a finite concave function on D.
Let $K := \{x \in D : 0 \le x_i \le 1, \forall i \in I\}$ and define $p(x) = \sum_{i \in I} x_i(1 - x_i)$.
Clearly, p is a concave function with nonnegative values on K and

$$\{x \in D,\ x_i \in \{0,1\}\} = \{x \in K : p(x) = 0\} = \{x \in K : p(x) \le 0\}.$$

Hence (12) can be reformulated as

$$\min \{f(x) : x \in K,\ p(x) \le 0\}$$

which is equivalent to, for any $t > t_o$

$$\min \{f(x) + tp(x) : x \in K\}. \tag{13}$$

A special case of the mixed zero-one concave minimization is linearly constrained quadratic zero-one programming which is known to be NP-hard and plays a key role in combinatorial optimization ([22]):

$$\alpha = \min \left\{ f(x) := \frac{1}{2}x^T Q x + q^T x : Ax \le b,\ x \in \{0,1\}^n \right\}, \tag{14}$$

where Q is an $(n \times n)$ symmetric matrix, q, $x \in \mathbb{R}^n$, A is an $(m \times n)$ matrix and $b \in \mathbb{R}^m$.
Let $\Omega := \{x \in \mathbb{R}^n : Ax \le b,\ x \in \{0,1\}^n\}$, $K := \{x \in \mathbb{R}^n : Ax \le b,\ x \in [0,1]^n\}$.
Since $\langle x, x \rangle = \langle e, x \rangle$ for all x in Ω (e is the vector of ones in $\mathbb{R}^n$), for any real number $\bar{\lambda}$ we have

$$\frac{1}{2}x^T Q x + q^T x = \frac{1}{2}\langle x, (Q - \bar{\lambda}I)x \rangle + \langle q + \frac{\bar{\lambda}}{2}e, x \rangle,\ \forall x \in \Omega.$$

Now let $\bar{\lambda} \ge 0$ be such that the matrix $Q - \bar{\lambda}I$ is negative semidefinite and let $p(x) = (1/2)(\langle e, x \rangle - \langle x, x \rangle)$. We can then write (14) in the form

$$\min\{f(x) := \frac{1}{2}\langle x, (Q - \bar{\lambda}I)x \rangle + \langle q + \frac{\bar{\lambda}}{2}e, x \rangle : x \in K,\ p(x) \le 0\} \tag{15}$$

which is equivalent to the (strictly) concave quadratic minimization program for any $t > t_o$: $(\lambda = \bar{\lambda} + t)$

$$\min \left\{ \frac{1}{2}\langle x, (Q - \lambda I)x \rangle + \langle q + \frac{\lambda}{2}e, x \rangle : x \in K \right\}. \tag{16}$$

Remark 3 *The equivalence between (14) and (16) is not unusual. Neverthe-less the existing algorithms for solving the last problem are not often applied in the solution of the former. The reason is that, in practice a branch-and-bound scheme (in a continuous approach) is often used for finding an ε-solution of the problem being considered, since it is very difficult to obtain an exact solu-tion. Unfortunately, such an ε-solution of (16) may be not integer. However, as has been shown in Theorem 2, the DCA applied to (16) with t sufficiently large converges to an integer feasible solution. This advantage of the DCA makes our method efficient in finding an integer ε-solution of (15) in the large scale setting.*

Using d.c. decomposition (6) for (16) we have

$$g(x) = \chi_K(x); \quad h(x) = -\left[\frac{1}{2}\langle x, (Q - \lambda I)x + \langle q + \frac{\lambda}{2}e, x\rangle\right], \quad (17)$$

and then the corresponding DCA scheme (7) can be described as follows:
DCA1.   Let $\varepsilon > 0$ be small enough and $x^0 \in \mathbb{R}^n$. Set $k \leftarrow 0, er \leftarrow 1$.
   While $er > \varepsilon$ do
      $y^k \leftarrow -(Q - \lambda I)x^k - q - (\lambda/2)e$.
      Solve the following linear program to obtain x^{k+1}:

$$\max\{\langle x, y^k\rangle : x \in K\}$$

   $er \leftarrow \|x^{k+1} - x^k\|$, $k \leftarrow k+1$
   end while
When using the branch and bound method we consider another d.c. decom-position of f with the separable concave part

$$g(x) := \frac{1}{2}x^T(Q + (\rho - \lambda)I)x + (q + \frac{\lambda}{2}e)^T x + \chi_K(x), \quad h(x) := \frac{1}{2}\rho \sum_{j=1}^{n} x_j^2, \quad (18)$$

ρ is a positive number such that the matrix $(\rho - \lambda)I + Q$ is positive semidefinite. We can then compute lower bounds as shown in Section 3.2.

4.2   Concave Minimization Programs over Efficient Sets

Let K be a bounded polyhedral convex set in $\mathbb{R}^n$ and C be a $(r \times n)$ - real matrix. Consider the following multiple objective linear programming problem

$$(MP) \qquad \max\{Cx : x \in K\}.$$

A point x is said to be *efficient* or *Pareto* for Problem (MP) if there does not exists $y \in K$ such that $Cy \geq Cx$, $Cy \neq Cx$. We denote by $E(K)$ the set of all efficient points of Problem (MP). This set is a union of faces of K [21], and in general it is not convex.

    Assume that $E(K)$ is nonempty. Concave minimization over efficient sets can be written as

$$\alpha := \min\{f(x) : x \in E(K)\}. \quad (19)$$

This problem has many applications in multiple objective decision making and recently has been extensively studied (see e.g., [7], [8], [4], [17] and their references). Suppose that the set K is expressed as $K := \{x \in \mathbb{R}^n : Ax \leq b, x \geq 0\}$. We consider the function p defined by, for each $x \in K$ ([4])

$$p(x) = \max\left\{e^T C(y - x) : Cy \geq Cx, y \in K\right\},$$

and extend p to the whole space by setting $p(x) = -\infty$ if $x \notin K$.

By writing

$$-p(x) := \min_y \left\{e^T C(x - y) : Cy \geq Cx, Ay \leq b, y \geq 0\right\} \qquad (20)$$

and setting the dual program of (20) we obtain

$$-p(x) = \langle C^T e, x\rangle + \max_{u,v}\left\{\langle Cx, u\rangle - \langle b, v\rangle : C^T u - A^T v \leq -C^T e, u \geq 0, v \geq 0\right\}. \qquad (21)$$

Hence $-p$ is convex on $\mathbb{R}^n$, and dom p is the projection onto $\mathbb{R}^n$ of $\{(x,y) \in \mathbb{R}^n \times K : Cy \geq Cx\}$. On the other hand, $p(x) \geq 0$, $\forall x \in K$. Furthermore $x \in E(K)$ if and only if $x \in K$ and $p(x) = 0$.

Finally, (19) can be written in the form

$$\min\{f(x) : p(x) \leq 0, x \in K\} \qquad (22)$$

which is equivalent to (P_t) for a sufficiently large number t.

We now apply the DCA scheme (7) for solving (19) when f is a concave quadratic function, say $f(x) := \frac{1}{2}\langle x, Qx\rangle + \langle x, q\rangle$ with Q being a $n \times n$-negative semidefinite matrix and $q \in \mathbb{R}^n$.

From (21) we see that a subgradient of $-p$ at a point $x \in K$ can be computed by solving one linear program ([9]). The related DCA scheme (7) now becomes:

DCA2. Let $\varepsilon > 0$ be small enough and $x^0 \in K$. Set $k \leftarrow 0, er \leftarrow 1$

While $er > \varepsilon$ do

Solve the linear program

$$\max_{u,v}\left\{\langle Cx^k, u\rangle - \langle b, v\rangle : C^T u - A^T v \leq -C^T e, u \geq 0, v \geq 0\right\}$$

to obtain (u^k, v^k) and set $y^k \leftarrow -Qx^k - q + tC^T e + tC^T u^k$.

Solve the following linear program to obtain x^{k+1}:

$$(LPP) \qquad \max\{\langle x, y^k\rangle : x \in K\}.$$

$er \leftarrow \|x^{k+1} - x^k\|, k \leftarrow k + 1$

end while

4.3   Bilevel Programming

We consider the bilevel program in which the objective function of the upper level problem is concave while the one of the lower level problem is linear, and both upper and lower level constraints are linear. More precisely, the problem takes the form

$$
\begin{cases}
\min f(x,y) \\
\text{s.t. } x \in K,\ A_1 x + B_1 y \leq b_1 \text{ and } y \text{ solves the linear program} \\
\min\{d^T y : y \in L, A_2 x + B_2 y \leq b_2\} \quad \text{(LP)}
\end{cases}
\tag{23}
$$

where f is a concave function on $\mathbb{R}^n \times \mathbb{R}^m$, A_1 (resp. A_2) is a $r \times n$ (resp. $q \times n$) real matrix, B_1 (resp. B_2) is a $r \times m$ (resp. $q \times m$) real matrix, $d \in \mathbb{R}^m$, $b_1 \in \mathbb{R}^r$, $b_2 \in \mathbb{R}^q$. The sets K and L place additional restrictions on the variables, and can be expressed as $K := \{x \in \mathbb{R}^n : A_0 x \leq a_0\}$, $L := \{y \in \mathbb{R}^m : B_0 y \leq b_0\}$, where A_0 (resp. B_0) is a $l \times n$ (resp. $s \times m$) real matrix, $a_0 \in \mathbb{R}^l$, $b_0 \in \mathbb{R}^s$. They are supposed to be nonempty and bounded. Furthermore we assume that the feasible set of Problem (23) is nonempty.

Let $C = \{(x,y) \in K \times L : A_1 x + B_1 y \leq b_1, A_2 x + B_2 y \leq b_2\}$. For each $x \in \mathbb{R}^n$ such that the set $\{y \in L : A_2 x + B_2 y \leq b_2\} \neq \emptyset$ we define the function p_1 by

$$
p_1(x) := \min\{d^T y : y \in L, A_2 x + B_2 y \leq b_2\}.
\tag{24}
$$

We extend p_1 to the whole space by setting $p_1(x) = +\infty$ when $\{y \in L : A_2 x + B_2 y \leq b_2\} = \emptyset$.

By using the Lagrangian duality in linear programming we get

$$
p_1(x) = \max_{\lambda, \mu} \left\{ \langle A_2 x - b_2, \lambda \rangle - \langle b_0, \mu \rangle : B_2^T \lambda + B_0^T \mu + d = 0, \lambda \geq 0, \mu \geq 0 \right\}.
\tag{25}
$$

Hence the function p_1 is convex on $\mathbb{R}^n$. Consider now the function $p(x,y)$ defined on $\mathbb{R}^n \times \mathbb{R}^m$ by

$$
p(x,y) = d^T y - p_1(x), \quad \forall (x,y) \in \mathbb{R}^n \times \mathbb{R}^m.
$$

It is clear that p is concave on $\mathbb{R}^n \times \mathbb{R}^m$, finite and nonnegative on C. Moreover, $x \in K \cap \{x : A_1 x + B_1 y \leq b_1\}$ and y solves the linear program (LP) if and only if $(x,y) \in C$ and $p(x,y) = 0$. Consequently, the bilevel program (23) takes the form of (P)

$$
\min \{f(x,y) : (x,y) \in C, p(x,y) \leq 0\}
\tag{26}
$$

which is equivalent to (for $t > t_0$)

$$
\min \{f(x,y) + tp(x,y) : (x,y) \in C\}.
\tag{27}
$$

We will use the DCA scheme (7) for solving the bilevel linear program, i.e. Problem (23) with $f(x,y) := a^T x + b^T y$. This case is related to the economy concept due to von Stackelberg (see [12] and references therein). Here the d.c. decomposition (6) is of the form $f(x,y) = g(x,y) - h(x,y)$ with

$$
g(x,y) = \chi_C(x,y),\ h(x,y) := -a^T x + tp_1(x) - (b+td)^T y.
$$

Hence the related DCA consists of computing, at each iteration k, w^k and (x^{k+1}, y^{k+1}) given by

$$w^k = -a + t\xi^k, (x^{k+1}, y^{k+1}) \in \arg \min_{(x,y) \in C} \{\langle (x,y), (w^k, b+td) \rangle\},$$

with $\xi^k \in \partial(p_1(x^k))$. From (25) we see that computing ξ^k amounts to solving a linear program of the form (25).

Finally the algorithm can be stated as follows:

DCA3. Let $\varepsilon > 0$, and $(x^0, y^0) \in dom\ \partial p$ be given. Set $k \leftarrow 0, er \leftarrow 1$
While $er > \varepsilon$ do
   Solve the next linear program to obtain (λ^k, μ^k):

$$\max_{\lambda, \mu} \left\{ \langle A_2 x^k - b_2, \lambda \rangle - \langle b_0, \mu \rangle : B_2^T \lambda + B_0^T \mu + d = 0, \lambda \geq 0, \mu \geq 0 \right\}.$$

Set $\xi^k := A_2^T \lambda^k$, and $w^k = -a + t\xi^k$.
Compute (x^{k+1}, y^{k+1}), a solution of the linear program

$$\min\{\langle (x,y), (w^k, b+td) \rangle : (x,y) \in C\}.$$

$er \leftarrow \|x^{k+1} - x^k\|, k \leftarrow k + 1$
end while

Remark 4 *If f is a concave quadratic function, say $f(x,y) = \frac{1}{2}x^T P x + a^T x + \frac{1}{2}y^T Q y + b^T y$, where P and D are symmetric negative semi-definite matrices, then the algorithm is the same as in DCA3 with a little modification: in each iteration k we have to compute (w^k, z^k) and (x^{k+1}, y^{k+1}) such that $w^k = -P x^k - a + t\xi^k, z^k = -Q y^k - b - td, (x^{k+1}, y^{k+1}) \in \arg\min_{(x,y) \in C}\{-\langle (x,y), (w^k, z^k) \rangle\}.$*

4.4   Concave Minimization
with Mixed Linear Complementarity Constraints

Let A be a $n \times n$ real matrix and let $b, c \in \mathbb{R}^n$. Let I be a given subset of $\{1, \dots, n\}$. The problem is defined by

$$\min \left\{ f(x) : Ax \leq b, x \geq 0, \sum_{i \in I} x_i(b - Ax)_i = 0 \right\} \tag{28}$$

where f is a finite concave function on the set $K := \{x \in \mathbb{R}^n : Ax \leq b, x \geq 0\}$ which is assumed to be bounded. Let $F_i(x) := b_i - A_i x$ (A_i denotes $i - th$ row of A) and define $p(x) = \sum_{i \in I} \min\{F_i(x), x_i\}$. Obviously, p is finite concave on $\mathbb{R}^n$ and p is nonnegative on K. Therefore, (28) can be written as

$$\min \{f(x) : x \in K, p(x) \leq 0\} \tag{29}$$

which is equivalent to (for $t > t_o$)

$$\min \left\{ f(x) + t \sum_{i \in I} \min\{F_i(x), x_i\} : x \in K \right\}. \tag{30}$$

We develop now the DCA scheme (7) for solving concave quadratic programs with mixed complementarity constraints, namely Problem (30) when $f(x) := \frac{1}{2}x^T P x + d^T x$, where P is symmetric negative semidefinite. With $g(x) := \chi_K(x), h(x) := -\frac{1}{2}x^T P x - d^T x - t \sum_{i \in I} \min\{b_i - A_i x, x_i\}$ (7) can be stated as: compute y^k and x^{k+1} such that

$$y^k = -Px^k - d + t\xi^k, x^{k+1} \in \ \arg\min_{x \in K}\{- \langle x, y^k \rangle\}$$
$$\text{with } \xi^k \in \partial \left(- \sum_{i \in I} \min\{F_i(x), x_i\}\right).$$

We can compute ξ^k explicitly as shown below. The algorithm is quite simple, its consists of solving one linear program at each iteration: (denote by e_i the vector in $\mathbb{R}^n$ which equal to one at ith component, zero otherwise).

DCA4. Let $\varepsilon > 0$, and $x^0 \in \mathbb{R}^n$ be given. Set $k \leftarrow 0, er \leftarrow 1$
While $er > \varepsilon$ do
For all $i \in I$ set $\varsigma_i^k \in \partial \left(\max(-(b_i - A_i x), -x_i)\right)$ as follows:

$\varsigma_i^k = -A_i^T$ if $x_i > b_i - A_i x$, $-e_i$ if $x_i < b_i - A_i x$, $\varsigma_i^k \in \left[-e_i, -A_i^T\right]$ if $x_i = b_i - A_i x$.

Set $\xi^k = \sum_{i \in I} \varsigma_i^k$, and $y^k = -Px^k - d + t\xi^k$.
Compute x^{k+1}, a solution of the linear program

$$\max_{x \in K} \langle x, y^k \rangle.$$

$er \leftarrow \|x^{k+1} - x^k\|, k \leftarrow k + 1$
end while

5   Conclusion

We have presented a continuous approach for a class of difficult nonconvex problems. The idea is based on the reformulation via exact penalty to eliminate the reverse convex constraint and transforms these problems into d.c. programs. This is motivated by the universality of d.c. programming and the success of DCA in global optimization which has been recently extensively developed in the literature. The DCA is quite special attractive because it works in a continuous domain of (P_t) but it provides a feasible solution of (P) which may be discrete (in quadratic zero one programming for example). The four algorithms presented above are quite simple and easy to implement: they require only linear programs as subproblems. Implementations of these algorithms are currently being done, except for quadratic zero one programs for which DCA and the combined algorithm developed in [3] have been shown to be very efficient already (DCA provided a global solution in several cases, see [3]). On the other hand, we plan to investigate good procedures of lower bounding to make efficient the combined DCA - branch and bound algorithms. Work in this direction is in progress.

References

[1] LE THI HOAI AN, *Contribution à l'optimisation non convexe et l'optimisation globale: Théorie, Algorithmes et Applications*, Habilitation à Diriger des Recherches, Université de Rouen, Juin 1997.

[2] LE THI HOAI AN and PHAM DINH TAO, Solving a class of linearly constrained indefinite quadratic problems by D.c. algorithms, *Journal of Global Optimization,* **11** (1997), pp. 253-285.

[3] LE THI HOAI AN and PHAM DINH TAO, A continuous approach for large-scale linearly constrained quadratic zero-one programming, *Optimization*, Vol. 50, No. 1-2, pp. 93-120 (2001).

[4] LE THI HOAI AN, PHAM DINH TAO and LE DUNG MUU, Numerical solution for Optimization over the efficient set by D.c. Optimization Algorithm, *Operations Research Letters*, 19 (1996) pp. 117-128.

[5] LE THI HOAI AN, PHAM DINH TAO and LE DUNG MUU, Exact penalty in d.c. programming, *Vietnam Journal of Mathematics*, 27:2 (1999), pp. 169-178.

[6] LE THI HOAI AN, PHAM DINH TAO and LE DUNG MUU, Simplicially constrained d.c. Optimization for optimizing over the Efficient and weakly efficient sets, *Journal of Optimization Theory and Applications, Vol 117, No 3 (2003), pp 503-531.*

[7] BENSON, H. P., Optimization over the Efficient Set, *Journal of Mathematical Analysis and Applications*, Vol.98, pp. 562-580, 1984.

[8] BOLITINEANU, S., Minimization of a Quasi-concave Function over an Efficient Set, *Mathematical Programming*, Vol. 61, pp. 89-110, 1993.

[9] J. B. HIRIAT URRUTY and C. LEMARECHAL, *Convex Analysis and Minimization Algorithms*, Springer Verlag Berlin Heidelberg, 1993.

[10] R. HORST, T. Q. PHONG, N. V. THOAI and J. VRIES, On Solving a D. C. Programming Problem by a Sequence of Linear Programs, *Journal of Global Optimization* 1(1991), pp. 183-203.

[11] R. HORST and H. TUY, *Global optimization (Deterministic approaches)*, Springer-Verlag, Berlin (1993).

[12] P. LORIDAN, J. MORGAN, Approximation solutions for two-level optimization problems, *Trends in Mathematical Optimization, International Series of Numer. Math.* Vol 84 (1988), Birkhäuser, pp. 181-196.

[13] PHAM DINH TAO, Algorithms for solving a class of non convex optimization problems. Methods of subgradients,*Fermat days 85. Mathematics for Optimization*, J. B.Hiriart Urruty (ed.), Elsevier Science Publishers B. V. North-Holland, 1986.

[14] PHAM DINH TAO, S.EL. BERNOUSSI, Duality in d.c. (difference of convex functions) optimization. Subgradient methods, *Trends in Mathematical Optimization, International Series of Numer Math.* Vol 84 (1988), Birkhauser, pp. 277-293.

[15] PHAM DINH TAO and LE THI HOAI AN, D.c. optimization algorithms for solving the trust region subproblem, *SIAM J. Optimization*, Vol 8, No 2 (1998), pp. 476-505.

[16] PHAM DINH TAO and LE THI HOAI AN, Convex analysis approach to d.c. programming. Theory, algorithms and applications, (Dedicated to Professor Hoang Tuy on the occasion of his 70th birth day), *Acta Mathematica Vietnamica*, Vol. 22, Number 1 (1997), pp. 289-355.

[17] J. PHILIP, Algorithms for the vector maximization problem, *Mathematical Programming* 2 (1972), pp. 207-229.

[18] ROCKAFELLAR, R. T., *Convex Analysis*. Princeton University Press, 1970.

[19] K. SHIMIZU, E. AIYOSHI, A new computation method for Stackelberg and min-max problem by use of a penalty method, *IEEE Transactions on Automatic Control*, Vol. AC-26, No. 2(1981), pp. 460-466.

[20] M. SIMAAN, J. CRUZ, On the Stackelberg strategy in nonzero-sum games, *Journal of Optimization Theory and Applications* 11, No. 5 (1973), pp. 533-555.

[21] STEUER R. E., *Multiple Criteria Optimization: Theory, Computation and Application*, John Willey and Sons, New York, New York 1986.

[22] S. A. VAVASIS (1991), *Nonlinear Optimization, Complexity Issues*, Oxford University Press.

[23] P. T.THACH, D.c. sets, d.c. functions and nonlinear equations, *Mathematical Programming*, 58 (1993), pp 415-428.

[24] H. TUY, A general deterministic approach to global optimization via d.c. programming, *Fermat days 85, Mathematics for Optimization*, J. B.Hiriart Urruty (ed.), Elsevier Science Publishers B. V. North-Holland (1986), pp. 273-303.

[25] H. TUY, Global minimization of difference of two convex functions, *Mathematical Programming Study* 30 (1987), pp. 159-182.

[26] H. TUY, D.c. programming, *In Handbook of Global Optimization, edited by R. Horst and Panos M. Pardalos, Kluwer Academic Publishers (1995)*.

[27] H. TUY, *Convex Analysis and Global Optimization*, Kluwer Academic Publishers, Boston, Dordrecht, London, 1998.

Symbolic-Interval Heuristic
for Bound-Constrained Minimization*

Evgueni Petrov

IRIN — Université de Nantes
2 rue de la Houssinière BP 92208
44322 Nantes Cedex 03 France
evgueni.petrov@irin.univ-nantes.fr

Abstract. Global optimization subject to bound constraints helps answer many practical questions in chemistry, molecular biology, economics. Most of algorithms for solution of global optimization problems are a combination of interval methods and exhuastive search. The efficiency of such algorithms is characterized by their ability to detect and eliminate sub-optimal feasible regions. This ability is increased by availability of a good upper bound on the global minimum. In this paper, we present a symbolic-interval algorithm for calculation of upper bounds in bound-constrained global minimization problems and report the results of some experiments.

1   Introduction

Global optimization subject to bound constraints helps answer many practical questions in chemistry, molecular biology, economics. Most of algorithms for solution of global optimization problems are a combination of interval methods and exhuastive search [3, 5, 9]. The efficiency of such algorithms is characterized by their ability to detect and eliminate sub-optimal feasible regions. This ability is increased by availability of a good upper bound on the global minimum.

A non-trivial upper bound on the minimum of any rational function can be calculated using modal intervals or Kaucher arithmetic [8, 1]. As far as non-linear functions are concerned in general, there is no similar means at present and one has to use some ad hoc combination of sampling and local search. The advantage of this approach is that it is applicable to any function specified by a black box transforming arguments into values. Its most significant disadvantage is that it ignores the symbolic representation of the minimized function which carries the helpful information on the global behaviour of the function.

In the this paper, we present a symbolic-interval algorithm for calculation of upper bounds in bound-constrained global minimization problems and report the results of some experiments.

---

* Financially supported by Centre Franco-Russe Liapunov (Project 06–98), by European project COCONUT IST–2000–26063.

The paper is structured as follows. Section 2 introduces the basic definitions. Section 3 presents the algorithm. In Section 4 we report the data of experiments with the Dixon-Szegö and some other classical functions. Section 5 concludes the paper. Appendix A contains the test functions.

2 Basic Definitions, Notation

This section contains the definitions linking real and interval functions, their symbolic representation and partial derivatives.

Intervals are closed convex sets of real numbers, boxes are interval vectors. The set of real numbers and the set of intervals are denoted by $\mathbb{R}$, $\mathbb{I}$. Interval function $[f]$ which returns the exact interval bound on the values of real function f given interval bounds on each of its real arguments is called *interval extension* of f. We call addition, subtraction, multiplication, division, power function for integer exponent, sine, cosine, tangent, exponent function and their inverses *basic functions*. We write "vector", "function" instead of "real vector", "real function".

The symbols $\mathcal{B}$ which denote the basic functions are called *function symbols*. The symbols of the arithmetic operations are *binary*, the others are *unary*. The symbols which denote real numbers are called *constant symbols*. The symbols v_i's distinct from the constant and function symbols are called *variables*.

The set of *terms* over the variables v_i's built from the symbols $\mathcal{B}$ and the constant symbols is defined as usual. A term t' is a *subterm* of a term t, or $t' \sqsubseteq t$ in symbols, if t' occurs in t. The term built from a symbol $\alpha \in \mathcal{B}$ and terms t', t'' is written as $\alpha(t', t'')$ (for binary α) or $\alpha(t')$ (for unary α). The set of variables in t is denoted by VAR(t).

Consider the following algebras having the same function symbols $\alpha \in \mathcal{B}$: terms $\mathcal{T}_n$ over the variables v_1, ..., v_n where (the operation denoted by) α returns the term built from the arguments (of α) and symbol α; functions $\mathcal{R}_n$ of n-dimensional vectors where α returns the composition of the arguments and the basic function denoted by α; interval functions $\mathcal{I}_n$ of n-dimensional boxes where α returns the composition of the arguments and the interval extension of the basic function denoted by α.

Algebra $\mathcal{T}_n$ is called the algebra of *real terms*. The subalgebra of $\mathcal{R}_n$ generated by the projections from $\mathbb{R}^n$ to $\mathbb{R}$ and the constant functions is called the algebra of *elementary functions*. The subalgebra of $\mathcal{I}_n$ generated by the projections from $\mathbb{I}^n$ to $\mathbb{I}$ and the functions returning singleton intervals is called the algebra of *interval elementary functions*.

Elementary and interval elementary functions are images of real terms under the homomorphisms $(\cdot)^{\mathcal{R}_n}$ and $[\cdot]_n$ which map the variables and the constant symbols to the generators of the respective algebra. We say that a term t *specifies* the elementary function $t^{\mathcal{R}_n}$ and interval elementary function $[t]_n$. In what follows, n is some fixed non-negative integer; we write $(\cdot)^{\mathcal{R}}$, $[\cdot]$ instead of $(\cdot)^{\mathcal{R}_n}$, $[\cdot]_n$.

The term α' specifies the derivative of the function specified by $\alpha(v_1)$ where symbol $\alpha \in \mathcal{B}$ is unary. The terms α_1', α_2' specify the partial derivatives of the function specified by $\alpha(v_1, v_2)$ where α is binary. The term obtained from some term t by simultaneous substitution of terms t_i's for the variables v_i's is denoted by $t(t_1, \ldots, t_n)$. In the case where $\mathrm{VAR}(t)$ is $\{v_1, v_2\}$ or $\{v_1\}$, we use the abbreviations $t(t_1, t_2)$, $t(t_1)$.

The function $\partial t / \partial(\cdot)$ recursively defined on the subterms of term t as follows:

$$\partial t / \partial t = 1$$
$$\partial t / \partial t' = \alpha'(t') \cdot \partial t / \partial \alpha(t') \qquad \text{for some unary } \alpha \in \mathcal{B}, \text{ some } \alpha(t') \sqsubseteq t$$
$$\partial t / \partial t' = \alpha_1'(t', t'') \cdot \partial t / \partial \alpha(t', t'') \quad \substack{\text{for some binary } \alpha \in \mathcal{B}, \text{ some } \alpha(t', t'') \sqsubseteq \\ t}$$
$$\partial t / \partial t' = \alpha_2'(t'', t') \cdot \partial t / \partial \alpha(t'', t') \quad \substack{\text{for some binary } \alpha \in \mathcal{B}, \text{ some } \alpha(t'', t') \sqsubseteq \\ t}$$

is called *differentiation* of t. The term $\partial t / \partial t'$ is called *partial derivative* of t wrt $t' \sqsubseteq t$. We assume that the symbols $\cdot$ and 1 denote multiplication and the unit.

Suppose that t contains a single occurence of t'. Let $t = t''(v_1, \ldots, v_n, t')$ where $\mathrm{VAR}(t'') = \{v_1, \ldots, v_{n+1}\}$. The function $(\partial t / \partial t')^{\mathcal{R}} : \mathbb{R}^n \to \mathbb{R}$ is the composition of the partial derivative $D_{n+1}\left[(t'')^{\mathcal{R}_{n+1}}\right]$ wrt the last argument of the function $(t'')^{\mathcal{R}_{n+1}} : \mathbb{R}^{n+1} \to \mathbb{R}$ and the injection from $\mathbb{R}^n$ to $\mathbb{R}^{n+1}$ which sends every vector p to the vector $\left(p, (t')^{\mathcal{R}}(p)\right)$.

3   Heuristic for Bound-Constrained Minimization

The global minimum of any function can be bounded from above with the help of sampling and local search. Applying local search to a function that is non-convex a priori has two drawbacks: the risk of slow convergence and the risk of violation of bound constraints. These risks decrease as the point for starting local search approaches to the global minimizer. The goal for our algorithm is to find a point that satisfies the bound constraints and where the value of the objective is sufficiently small.

Outline of the algorithm. Let real term t, box b specify the objective, the bound constraints. Our algorithm repeatedly exploits the following observation concerning the minimum of the objective $t^{\mathcal{R}}$ in b. Let $t^{\mathcal{R}}$ satisfy $\forall p \in \mathbb{R}^n \ t^{\mathcal{R}}(p) = f(p, g(p))$ for some elementary functions f and g, and let a be the value of g at some point in b. Then the minimum of the objective is bounded from above by $\min_{p \in b \cap g^{-1}(a)} f(p, a)$ where $g^{-1}(a)$ is the pre-image of a under g.

The choice of a specific a is determined by the considerations of efficiency. The ideal (impractical) choice would be the value of g at the global minimizer of the objective. In practice, we demand that f be monotonic in the last argument and that mini- and maximization of g be affordable. Under these assumptions, we use $a = \min_b g$ (if f increases) or $a = \max_b g$ (if f decreases). The assumptions are valid, if g is specified by a subterm $t' \sqsubseteq t$ such that (1) no variable occurs

$p = (0, \ldots, 0)$
WHILE VAR(t) $\neq \emptyset$
    ℓ = THE SET OF GOOD SUBTERMS OF t
    IF $\ell = \emptyset$
        BREAK
    IF VAR(t') = VAR(t) FOR EACH $t' \in \ell$
        SET t' TO SUCH SUBTERM FROM ℓ THAT
            $t^{\mathcal{R}}(\text{PROJ}(t', \text{mid}[t'](b) - \text{sign}(\partial t/\partial t') \cdot \text{rad}[t'](b)))$ IS MINI-
MUM
    ELSE IF VAR(t') IS A SINGLETON SET FOR SOME $t' \in \ell$
        SET t' TO SUCH SUBTERM
    ELSE
        SET t' TO SOME SUBTERM FROM ℓ
    $p = p + \text{PROJ}(t', \text{mid}[t'](b) - \text{sign}(\partial t/\partial t') \cdot \text{rad}[t'](b))$
    SUBSTITUTE p_i FOR EACH $v_i \in \text{VAR}(t')$ IN t
SET p_i TO $\text{mid}(b_i)$ FOR EACH $v_i \in \text{VAR}(t)$

Fig. 1. The heuristic for bound-constrained minimization; vector p is the point for starting local search, set ℓ is the set of good subterms of t, term $t' \sqsubseteq t$ is some good subterm, function PROJ$(\cdot, \cdot)$ is defined in Fig. 2, function sign$(\cdot)$ takes a term specifying a function that does not change sign in b and returns its sign $(+1$ or $-1)$, functions mid$(\cdot)$ and rad$(\cdot)$ return the midpoint and the radius of intervals, interval b_i is the i-th element in b

in t' twice and (2) the function specified by the partial derivative $\partial t/\partial t'$ does not change sign in b. The maximal wrt $\sqsubseteq$ subterms of t that have these two properties are called *good subterms*. Our algorithm is provided in Fig. 1 and Fig. 2.

Example. Let us trace the algorithm for the Branin function specified by

$$t = \left(v_2 - \frac{5.1}{4\pi^2} v_1^2 + \frac{5v_1}{\pi} - 6 \right)^2 + 10 \left(1 - \frac{1}{8\pi} \right) \cos v_1 + 10$$

in the box $b = ([-5, 10], [0, 15])$ (we write t in the infix notation).

There is only one good subterm $t' = 10 \left(1 - \frac{1}{8\pi} \right) \cos v_1 \sqsubseteq t$ (one can check that $\partial t/\partial t' = 1 \cdot 1 \cdot 1$). Minimization of $(t')^{\mathcal{R}}$ in b assigns $-\pi$ to v_1 (or π, or 3π depending on the actual implementation of PROJ$(\cdot, \cdot)$). Since VAR(t) = $\{v_2\} \neq \emptyset$, the body of the loop is run once again. Now the entire term t is its own good subterm (v_1 has been replaced with $-\pi$!). Minimization of $t^{\mathcal{R}}$ assigns 12.275 to v_2 and the loop terminates for VAR(t) is now $\emptyset$. The point p for starting local search is $(-\pi, 12.275)$. Actually, it is one of the global minimizers of the Branin function in b.

Correctness. The algorithm in Fig. 1 and Fig. 2 is correct; that is it generates a vector p that satisfies each bound constraint.

$$\text{PROJ}(v_i, c) = (0, \ldots, 0, c, 0, \ldots, 0) \qquad \text{where } c \text{ takes the } i\text{-th place}$$
$$\text{PROJ}(\alpha(t'), c) = \text{PROJ}(t', c') \qquad \text{where } c = \alpha^{\mathcal{R}}(c'), \, c' \in [t'](b)$$
$$\text{PROJ}(\alpha(t', t''), c) = \text{PROJ}(t', c') + \text{PROJ}(t'', c'') \quad \begin{array}{l} \text{where } c = \alpha^{\mathcal{R}}(c', c''), \\ c' \in [t'](b), \, c'' \in [t''](b) \end{array}$$

Fig. 2. The definition of the function $\text{PROJ}(\cdot, \cdot)$; symbol v_i is a variable, c, c', c'' are real numbers, terms t', t'' are subterms of t, function $\alpha^{\mathcal{R}}$ is the basic function denoted by $\alpha \in \mathcal{B}$

Correctness of the algorithm follows from the fact that, for the terms t' where each variable occurs no more than once, the interval function $[t']$ is the interval extension of the function $(t')^{\mathcal{R}}$. In particular, it means that the bounds of $[t'](b)$ are the global extrema of $(t')^{\mathcal{R}}$ in b. Therefore the i-th element in the vector $\text{PROJ}(t', a)$ where a is any of the bounds of $[t'](b)$ belongs to b_i whenever $v_i \in \text{VAR}(t')$.

The worst case complexity. In the worst case, the number of evaluations of the basic functions and their interval extensions is $O(s^2/n)$, the number of comparisons is $O(n \cdot s)$ where s is the size of the term specifying the objective, n is the number of the variables.

Computer implementation details. The computer implementation of the algorithm in Fig. 1 and Fig. 2 must operate correctly on the real numbers non-representable in the floating point form. Instead of such real numbers, we use thin intervals bounded by floating point numbers. That is the computer implementation of our algorithm needs 4 floating numbers per interval.

The intervals $[\partial t/\partial t'](b)$ and $[t'](b)$ are calculated for all subterms $t' \sqsubseteq t$ by the interval version of automatic differentiation [4].

4   Experiments with the Dixon-Szegö and Some other Functions

In this section, we summarize the experiments with a C++ implementation of the algorithm described in Section 3. The functions for these experiments are taken from [2], [7] and the 1st International Contest on Evolutionary Optimization (see Appendix A).

For each test function, we report the number n of the variables in the corresponding term, the global minimum and the upper bound calculated by our heuristic (the 4 first digits from their decimal representation) (see Fig. 3). The general observation concerning the algorithm is that it tends to perform better on separable objective functions.

Dixon-Szegö functions				Janson-Knüppel functions (cntd.)			
function	n	min	upper b.	function	n	min	upper b.
Shekel 5	4	-10.38	-10.38	Levy 8 bis	2 through 80	0.000	0.000
Shekel 7	4	-10.48	-10.48	Levy 13	2 through 80	0.000	0.000
Shekel 10	4	-10.53	-10.53				
Goldstein-Price	2	3.000	600.0	Int. Contest on Evol. Optim. functions			
6 hump camel	2	-1.031	150.9	ICEO 2	5	0.000	0.000
Hartmann 3	3	-3.862	-3.761	ICEO 2	10	0.000	0.000
Hartmann 6	6	-3.322	-3.203	ICEO 3	5	-10.40	-10.40
Shubert	2	-186.7	19.87	ICEO 3	10	-10.20	-10.20
Branin	2	0.3978	0.3978	ICEO 4	5	-4.687	-4.488
Janson-Knüppel functions				ICEO 4	10	-9.660	-8.048
Rosenbrock	2 through 80	0.000	0.000	ICEO 5	5	-1.499	-1.499
Levy 8	2 through 80	0.000	0.000	ICEO 5	10	-1.500	-1.500

Fig. 3. The upper bounds found by the algorithm for the Dixon-Szegö, Janson-Knüppel and 1st International Constest on Evolutionary Optimization functions; each line contains the name of a function from Appendix A, the dimension(s) tried for this function, its global minimum, the calculated upper bound

5   Conclusion

In this paper we have presented a symbolic-interval heuristic for bound-constrained minimization problems. The key idea of our heuristic is to simplify the minimization problem by ignoring the dependencies between certain subexpressions in the symbolic representation of the objective. Our experiments indicate that this approach allows one to calculate good upper bounds in a number of bound-constrained minimization problems.

Our future work will focus on the theoretical properties of the presented heuristic as well as on its integration into modern algorithms for global optimization like [6].

References

[1] J. Armengol Llobet. Application of modal interval analysis to the simulation of the behaviour of dynamic systems with uncertain parameters. Ph. D. thesis, 1999.

[2] L. C. W. Dixon and G. P. Szegö, editors. *Towards Global Optimization*. North-Holland, Amsterdam, 1975–1978.

[3] A. V. Fiacco and G. P. McCormick. *Nonlinear Programming. Sequential unconstrained minimization techniques*. Willey, reprinted by SIAM publications, 1990.

[4] A. Griewank. On automatic differentiation. In M. Iri and K. Tanabe, editors, *Mathematical Programming: Recent Developments and Applications*, pages 83–108. Kluwer Academic Publishers, 1989.

[5] E. Hansen. *Global optimization using interval analysis*, volume 165 of *Pure and applied mathematics*. Marcel Dekker, 1992.

[6] W. Huyer and A. Neumaier. Global optimization by multilevel coordinate search. *Journal of Global Optimization*, 14:331–355, 1999.

[7] C. Jansson and O. Knüppel. A global minimization method: the multi-dimensional case. Technische Informatik-III, TU Hamburg-Harburg, 1992.

[8] E. Kaucher. Interval analysis in the extended interval space IR. *Computing*, Suppl. 2:33–49, 1980.

[9] J. Nocedal and S. J. Wright. *Numerical Optimization*. Springer Series in Operations Research. Springer, 1999.

A   Test Functions

For the sake of completeness we give the terms and the boxes that specify the test objective functions and the corresponding bound constraints. We denote the n-element vector $(x, \ldots, x)$ by x^n. The terms are written in the infix notation. The symbolic sums and products must be expanded into the appropriate chains of left-associative additions and multiplications.

Shekel m: $b = [0, 10]^4$, $t = -\sum_{i=1}^{m} 1 \Big/ \left(c_i + \sum_{j=1}^{4} (v_j - a_{ij})^2 \right)$

$$c = \left(1\ 2\ 2\ 4\ 4\ 6\ 3\ 7\ 5\ 5 \right) / 10, \quad a^{\mathrm{T}} = \begin{pmatrix} 4\ 1\ 8\ 6\ 3\ 2\ 5\ 8\ 6\ 7 \\ 4\ 1\ 8\ 6\ 7\ 9\ 5\ 1\ 2\ 3.6 \\ 4\ 1\ 8\ 6\ 3\ 2\ 3\ 8\ 6\ 7 \\ 4\ 1\ 8\ 6\ 7\ 9\ 3\ 1\ 2\ 3.6 \end{pmatrix}$$

Goldstein–Price: $b = [-2, 2]^2$,

$$t = \left(1 + (v_1 + v_2 + 1)^2 \cdot (19 - 14 \cdot v_1 + 3 \cdot v_1^2 - 14 \cdot v_2 + 6 \cdot v_1 \cdot v_2 + 3 \cdot v_2^2) \right) \times$$
$$\left(30 + (2 \cdot v_1 - 3 \cdot v_2)^2 \cdot (18 - 32 \cdot v_1 + 12 \cdot v_1^2 + 48 \cdot v_2 - 36 \cdot v_1 \cdot v_2 + 27 \cdot v_2^2) \right)$$

6 Hump Camel:

$$b = ([-3, 3], [-2, 2]), \quad t = \left(4 - 2.1 \cdot v_1^2 + v_1^4/3 \right) \cdot v_1^2 + v_1 \cdot v_2 + 4 \cdot \left(v_2^2 - 1 \right) \cdot v_2^2$$

Hartmann 3: $b = [0, 1]^3$, $t = -\sum_{j=1}^{4} c_j \cdot \exp \left(-\sum_{i=1}^{3} a_{ij} \cdot (v_i - p_{ij})^2 \right)$,

$$a = \begin{pmatrix} 3 & 0.1 & 3 & 0.1 \\ 10 & 10 & 10 & 10 \\ 30 & 35 & 30 & 35 \end{pmatrix}, \quad p = \begin{pmatrix} 0.36890 & 0.46990 & 0.10910 & 0.03815 \\ 0.11700 & 0.43870 & 0.87320 & 0.57430 \\ 0.26730 & 0.74700 & 0.55470 & 0.88280 \end{pmatrix},$$

$$c = \left(1\ 1.2\ 3\ 3.2 \right)$$

Hartmann 6: $b = [0, 1]^6$, $t = -\sum_{j=1}^{4} c_j \cdot \exp \left(-\sum_{i=1}^{6} a_{ij} \cdot (v_i - p_{ij})^2 \right)$,

$$a^{\mathrm{T}} = \begin{pmatrix} 10 & 3 & 17 & 3.5 & 1.7 & 8 \\ 0.05 & 10 & 17 & 0.1 & 8 & 14 \\ 3 & 3.5 & 1.7 & 10 & 17 & 8 \\ 17 & 8 & 0.05 & 10 & 0.1 & 14 \end{pmatrix}, p^{\mathrm{T}} = \begin{pmatrix} 0.1312 & 0.1696 & 0.5569 & 0.0124 & 0.8283 & 0.5886 \\ 0.2329 & 0.4135 & 0.8307 & 0.3736 & 0.1004 & 0.9991 \\ 0.2348 & 0.1451 & 0.3522 & 0.2883 & 0.3047 & 0.6650 \\ 0.4047 & 0.8828 & 0.8732 & 0.5743 & 0.1091 & 0.0381 \end{pmatrix},$$

$$c = \left(1\ 1.2\ 3\ 3.2 \right)$$

Shubert: $b = [-10, 10]^2$, $t = \prod_{j=1}^{2} \sum_{i=1}^{6} i \cdot \cos((i+1) \cdot v_j + i)$

Rosenbrock: $b = [-2.048, 2.048]^n$, $t = \sum_{i=1}^{n-1} 100 \cdot (v_i - v_{i+1}^2)^2 + (1 - v_{i+1})^2$

Levy 8: $b = [-10, 10]^n$,

$$t = 10 \cdot \sin^2 \pi \cdot \frac{v_1 + 3}{4} + \left(\frac{v_n - 1}{4}\right)^2 + \sum_{i=1}^{n-1} \left(\frac{v_i - 1}{4}\right)^2 \cdot \left(1 + 10 \cdot \sin^2 \pi \cdot \frac{v_{i+1} + 3}{4}\right)$$

Levy 8 bis:

$$b = [-10, 10]^n, \quad t = 10 \cdot \sin^2 \pi \cdot v_1 + (v_n - 1)^2 + \sum_{i=1}^{n-1} (v_i - 1)^2 \cdot (1 + 10 \cdot \sin^2 \pi \cdot v_{i+1})$$

Levy 13: $b = [-10, 10]^n$,

$$t = \sin^2 3 \cdot \pi \cdot v_1 + (v_n - 1)^2 \cdot (1 + \sin^2 3 \cdot \pi \cdot v_n) + \sum_{i=1}^{n-1} (v_i - 1)^2 \cdot (1 + \sin^2 3 \cdot \pi \cdot v_{i+1})$$

ICEO 2: $b = [-600, 600]^n$, $t = \sum_{i=1}^{n} (v_i - 100)^2 / 4000 - \prod_{i=1}^{n} \cos\left((v_i - 100)/\sqrt{i}\right) + 1$

ICEO 3: $b = [0, 10]^n$, $t = -\sum_{j=1}^{30} 1/(c_j + \sum_{i=1}^{n} (v_i - a_{ji})^2)$ (see the matrix a and the vector c below)

ICEO 4: $b = [0, \pi]^n$, $t = -\sum_{i=1}^{n} \sin v_i \cdot \sin^{20}(i \cdot v_i^2 / \pi)$

ICEO 5: See the matrix a and the vector c below.

$$b = [0, 10]^n, \quad t = -\sum_{j=1}^{30} c_j \cdot \exp\left(-\sum_{i=1}^{n} (a_{ji} - v_i)^2 / \pi\right) \cdot \cos\left(\pi \cdot \sum_{i=1}^{n} (a_{ji} - v_i)^2\right)$$

The coefficients for ICEO 3 and ICEO 5:

$$
a = \begin{pmatrix}
9.681 & 0.667 & 4.783 & 9.095 & 3.517 & 9.325 & 6.544 & 0.211 & 5.122 & 2.020 \\
9.400 & 2.041 & 3.788 & 7.931 & 2.882 & 2.672 & 3.568 & 1.284 & 7.033 & 7.374 \\
8.025 & 9.152 & 5.114 & 7.621 & 4.564 & 4.711 & 2.996 & 6.126 & 0.734 & 4.982 \\
2.196 & 0.415 & 5.649 & 6.979 & 9.510 & 9.166 & 6.304 & 6.054 & 9.377 & 1.426 \\
8.074 & 8.777 & 3.467 & 1.863 & 6.708 & 6.349 & 4.534 & 0.276 & 7.633 & 1.567 \\
7.650 & 5.658 & 0.720 & 2.764 & 3.278 & 5.283 & 7.474 & 6.274 & 1.409 & 8.208 \\
1.256 & 3.605 & 8.623 & 6.905 & 0.584 & 8.133 & 6.071 & 6.888 & 4.187 & 5.448 \\
8.314 & 2.261 & 4.224 & 1.781 & 4.124 & 0.932 & 8.129 & 8.658 & 1.208 & 5.762 \\
0.226 & 8.858 & 1.420 & 0.945 & 1.622 & 4.698 & 6.228 & 9.096 & 0.972 & 7.637 \\
7.305 & 2.228 & 1.242 & 5.928 & 9.133 & 1.826 & 4.060 & 5.204 & 8.713 & 8.247 \\
0.652 & 7.027 & 0.508 & 4.876 & 8.807 & 4.632 & 5.808 & 6.937 & 3.291 & 7.016 \\
2.699 & 3.516 & 5.874 & 4.119 & 4.461 & 7.496 & 8.817 & 0.690 & 6.593 & 9.789 \\
8.327 & 3.897 & 2.017 & 9.570 & 9.825 & 1.150 & 1.395 & 3.885 & 6.354 & 0.109 \\
2.132 & 7.006 & 7.136 & 2.641 & 1.882 & 5.943 & 7.273 & 7.691 & 2.880 & 0.564 \\
4.707 & 5.579 & 4.080 & 0.581 & 9.698 & 8.542 & 8.077 & 8.515 & 9.231 & 4.670 \\
8.304 & 7.559 & 8.567 & 0.322 & 7.128 & 8.392 & 1.472 & 8.524 & 2.277 & 7.826 \\
8.632 & 4.409 & 4.832 & 5.768 & 7.050 & 6.715 & 1.711 & 4.323 & 4.405 & 4.591 \\
4.887 & 9.112 & 0.170 & 8.967 & 9.693 & 9.867 & 7.508 & 7.770 & 8.382 & 6.740 \\
2.440 & 6.686 & 4.299 & 1.007 & 7.008 & 1.427 & 9.398 & 8.480 & 9.950 & 1.675 \\
6.306 & 8.583 & 6.084 & 1.138 & 4.350 & 3.134 & 7.853 & 6.061 & 7.457 & 2.258 \\
0.652 & 2.343 & 1.370 & 0.821 & 1.310 & 1.063 & 0.689 & 8.819 & 8.833 & 9.070 \\
5.558 & 1.272 & 5.756 & 9.857 & 2.279 & 2.764 & 1.284 & 1.677 & 1.244 & 1.234 \\
3.352 & 7.549 & 9.817 & 9.437 & 8.687 & 4.167 & 2.570 & 6.540 & 0.228 & 0.027 \\
8.798 & 0.880 & 2.370 & 0.168 & 1.701 & 3.680 & 1.231 & 2.390 & 2.499 & 0.064 \\
1.460 & 8.057 & 1.336 & 7.217 & 7.914 & 3.615 & 9.981 & 9.198 & 5.292 & 1.224 \\
0.432 & 8.645 & 8.774 & 0.249 & 8.081 & 7.461 & 4.416 & 0.652 & 4.002 & 4.644 \\
0.679 & 2.800 & 5.523 & 3.049 & 2.968 & 7.225 & 6.730 & 4.199 & 9.614 & 9.229 \\
4.263 & 1.074 & 7.286 & 5.599 & 8.291 & 5.200 & 9.214 & 8.272 & 4.398 & 4.506 \\
9.496 & 4.830 & 3.150 & 8.270 & 5.079 & 1.231 & 5.731 & 9.494 & 1.883 & 9.732 \\
4.138 & 2.562 & 2.532 & 9.661 & 5.611 & 5.500 & 6.886 & 2.341 & 9.699 & 6.500
\end{pmatrix}
\qquad
c = \begin{pmatrix}
0.806 \\
0.517 \\
0.1 \\
0.908 \\
0.965 \\
0.669 \\
0.524 \\
0.902 \\
0.531 \\
0.876 \\
0.462 \\
0.491 \\
0.463 \\
0.714 \\
0.352 \\
0.869 \\
0.813 \\
0.811 \\
0.828 \\
0.964 \\
0.789 \\
0.360 \\
0.369 \\
0.992 \\
0.332 \\
0.817 \\
0.632 \\
0.883 \\
0.608 \\
0.326
\end{pmatrix}
$$

A Global Constrained Optimization Algorithm for Engine Calibration

Aurélien Schmied

[1] Renault Direction de la Recherche
TCR RUC 4 31
1 avenue du golf, 78288 Guyancourt Cedex, France
aurelien.schmied@renault.com
phone:01 34 95 93 95
[2] Université Paris XII
Centre de mathématiques

Abstract. The engine calibration procedure developed in Renault involves solving a constrained global optimization problem. This is a medium scale problem with a great number of linear and non linear constraints and containing a large number of local minima. Therefore, we use a strategy combining global and local optimization. We propose a new global stochastic optimization algorithm called Multistoch generating a fixed number of starting points for a local optimization procedure (see [2]). These set of points are called a grid. This algorithm although different from simulated annealing (see [3]) involves a gibbs measure. This measure is used to select a point in the grid. Around this selected point we generate a new candidate point which according to his value will modify the grid. After a description of the engine calibration problem some theorical and numerical results on the algorithm are presented.

Keywords: Global constrained optimization, engine calibration, stochastic algorithm

1 The Engine Calibration Problem

In this section, we will introduce the aim of the engine calibration problem. The goal of the optimization is to set the engine controls so that the fuel consumption is as low as possible over a given cycle while complying with emission limits and meeting some driveability constraints.

1.1 Tests Cycles and Emission Limits

A test cycle is a profile of speed versus time that the vehicule has to realise. It describes a characteristic use of a car on a plane road. On this profile the legislator imposes some limitations on certain exhaust emissions such as soot and nitrogen monoxyde, etc. These constraints must be respected and the consumption should be minimized on the test cycle and meet constraints on driveability. In order to calibrate the engine on a given cycle and a given vehicule we apply the following method.

C. Bliek et al. (Eds.): COCOS 2002, LNCS 2861, pp. 111–122, 2003.

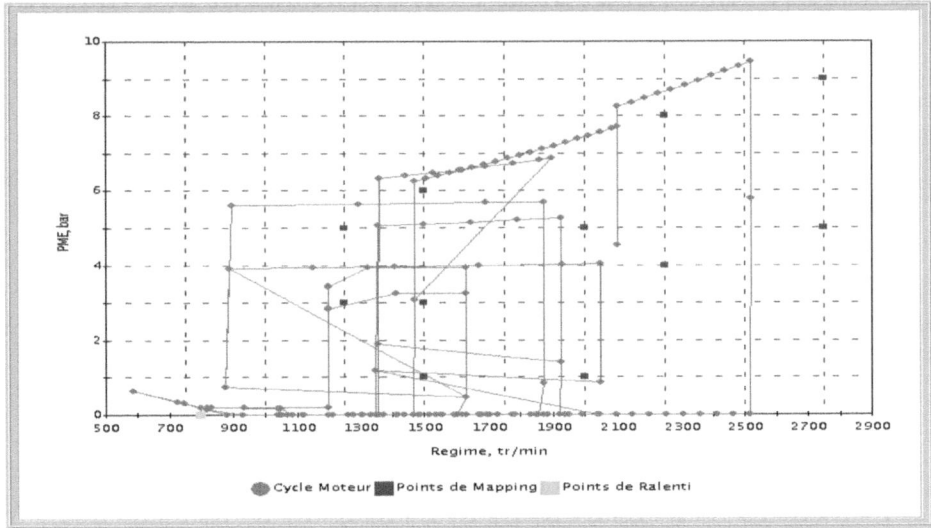

Fig. 1. Diagram of the European test cycle in the speed/load plane

1.2 The Engine Control

The test cycles are transposed for each vehicule in the speed and load plane of the engine. This plane represents the duty that is asked to the engine. The test cycle becomes a path (see figure 1) in the speed and load plane over which the engine has to meet the imposed limits on exhaust emissions and lower the fuel consumption.

In order to calibrate the engine, we choose a set of points (square points on figure 1) in the speed/load plane called operating points. They are weighted according to their relevancy to the cycle. This is a classical method for engine calibration (see [1]). These points are used for approximating the emissions on a given cycle and are also used to tune the engine. The control parameters to calibrate on each operating point are injection timing, quantity of fuel injected, pression at the admission, etc. The number of operating points determines the size of optimization problem. The problem formulation is the aim of the next section.

1.3 The Optimization Problem Formulation

The different functions used in the formulation are polynomials of maxi-mum fourth degree. These are statical models of the engine. Let us introduce:

- NP the number of operating points,
- $(X_1^p, \ldots, X_6^p)$ engine control parameters at the operating point p,
- $CONSU^p(X^p)$ represents the fuel consumption at the operating point p,
- $NOISE^p(X^p)$ represents the noise at the operating point p,

- $g_j^p(X^p)$ represents the j limited emission at the operating point p,
- limit$_j$ represents the limit on the j emission,
- limit$_{\text{NOISE}}$ represents the limit on the $NOISE$ emission,
- α_p represents the weight in the cycle for the operating point p,

We're looking for the control parameters $X = (X^1, \ldots, X^{NP}) \in D$, that minimize

$$\sum_{p=1}^{NP} \alpha_p CONSU^p(X^p),$$

under the constraints

$$\forall j \in \{1, \ldots, 6\}, \qquad \sum_{p=1}^{NP} \alpha_p g_j^p(X^p) \leq \text{limit}_j,$$

and

$$\forall p \in \{1, \ldots, NP\}, \qquad NOISE^p(X^p) \leq \text{limit}_{\text{NOISE}},$$

where $X^p = (X_1^p, \ldots, X_6^p)$ and D is a compact in $\mathbb{R}^d$ where $d = NP * 6$. The next section deals with solving the preceeding optimization problem.

2   Solving the Problem

Let us look at the characteristics of the engine calibration optimization problem.

2.1   Problem Characterisation

A typical engine calibration problem is with $NP = 15$ which gives 90 variables and more than a thousand local minima. Therefore, a strategy using global and local optimisations was developed. The aim was to find some good starting points for a local optimization (see [2]) which assured us of finding an optimal solution respecting the constraints in a limited time. Among these techniques are clustering methods such as Random Linkage (see [4]) which gather close points together so that a new local optimization is started from each cluster. Here, we propose a global stochastic algorithm called Multistoch which is based on a different heuristic which generates a fixed number of starting points for local optimization. The proposed technique is based on a set of points and that we move in D to explore the different local minima. This algorithm uses a general loop based on the Gibbs measure like simulated annealing (see [3]). Still simulated annealing is different as it tends to simulate a Gibbs measure. Next section gives the details of the algorithm.

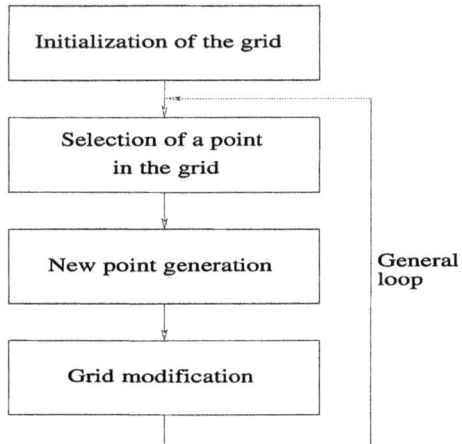

Fig. 2. A schematic view of the algorithm

2.2 Multistoch

The algorithm Multistoch's idea is to move a grid of N point throw a compact set $D \subset \mathbb{R}^d$. In the first place Multistoch selects Z_n a point of the grid Γ_{n-1} using a gibbs measure. The gibbs measure gives a probability measure for each point of the grid according to their objective function's value and a positive sequence $(T_n)_{n \in \mathbb{N}}$. Then, the algorithm generates a new candidate point around the selected point of the grid Z_n using a gaussian distribution. The gaussian distribution is centered in Z_n and has a large standard deviation $\rho > 0$ if the value of the objective function in Z_n is near the current minimum of the grid, otherwise it has a decreasing standard deviation $(\rho_n)_{n \in \mathbb{N}}$. Therefore, the new candidate point is generated near the selected point of the grid Z_n if it's value is interesting and in a larger area otherwise. Then the grid is modified according to the value of the objective function at the new candidate point.

Let us define f as the objective function which includes the penalized constraints and x^* it's argmin on D. In order to simplify the notations within the proofs we enlarge f over $\mathbb{R}^d$ by defining $f(x) = +\infty$ if x belongs to the complementary set of D in $\mathbb{R}^d$ written D^c.

Before defining the algorithm in detail, we introduce $(U_n, V_n)_{n \in \mathbb{N}}$ a sequence of independant random variables with common distribution belonging to $(0,1) \times \mathbb{R}^d$, defined over the probability space $(\Omega, \mathcal{A}, \mathcal{F})$ with U_1 and V_1 independant. U_1 has an uniform distribution over $[0,1]$ and V_1 has a normal distribution over $\mathbb{R}^d$. We initialize the algorithm with a grid Γ_0 of N points in D, possibly random and independant of $(U_n, V_n)_{n \in \mathbb{N}}$. We define $\mathcal{F}_n$ the σ-algebra generated by $(U_k, V_K, \Gamma_0^1, \dots, \Gamma_0^N, 1 \le k \le n)$. We consider the following algorithm defined with the preceeding notations.

- **Initialization**
  Let Γ_0 a grid of N points in D;
- **General loop**
  - **Selecting a point of the grid Γ_{n-1} with a Gibbs measure**
    We define Z_n a point of Γ_{n-1} such as

$$Z_n = \sum_{i=1}^{N} \Gamma_{n-1}^i \, \mathbb{1}_{\{U_n \in I_{n-1}^i\}}$$

where $I_{n-1}^i = \left[\dfrac{\exp(\frac{-f(\Gamma_{n-1}^{i-1})}{T_n})}{\sum_{j=1}^{N}\exp(\frac{-f(\Gamma_{n-1}^j)}{T_n})}, \dfrac{\exp(\frac{-f(\Gamma_{n-1}^i)}{T_n})}{\sum_{j=1}^{N}\exp(\frac{-f(\Gamma_{n-1}^j)}{T_n})} \right)$, $\quad \forall i = 2, \ldots, N;$

and $I_{n-1}^1 = \left[0, \dfrac{\exp(\frac{-f(\Gamma_{n-1}^1)}{T_n})}{\sum_{j=1}^{N}\exp(\frac{-f(\Gamma_{n-1}^j)}{T_n})} \right)$, where $\mathbb{1}_{\{A\}}(x) = 1$ if $x \in A$ and 0
elsewhere.
Therefore Z_n has for conditional law knowing $\mathcal{F}_{n-1}$

$$\mathbb{P}(Z_n = \Gamma_{n-1}^i | \mathcal{F}_{n-1}) = \frac{\exp(-f(\Gamma_{n-1}^i)/T_n)}{\sum_{j=1}^{N}\exp(-f(\Gamma_{n-1}^j)/T_n)}, \quad \forall i \in \{1, \ldots, N\},$$

where $\mathbb{P}$ stands for probability.
  - **Generation of a new candidate point Y_n**
    Let Y_n be a random variable defined by

$$Y_n = Z_n + V_n(\rho_n \mathbb{1}_{\{f(Z_n) < \min_i f(\Gamma_{n-1}^i) + \beta\}} + \rho \mathbb{1}_{\{f(Z_n) \geq \min_i f(\Gamma_{n-1}^i) + \beta\}}).$$

  - **Grid modification**
    We define

$$X_n = \left\{ \begin{array}{l} Y_n \text{ if } f(Y_n) < f(Z_n) \\ Z_n \text{ else} \end{array} \right\}$$

and

$$\Gamma_n = (\Gamma_{n-1} \backslash \{Z_n\}) \cup \{X_n\}.$$

In other words, we replace the selected point of the grid Z_n by the new candidate point Y_n only if $f(Y_n) < f(Z_n)$ otherwise, the grid renames unchanged.

Using a Gibbs measure allows when T_n is large enough to select Z_n uniformly over all the point of the grid, and when T_n is small enough to select only the point of the grid which have the smallest objective function value. We can notice that as f equals $+\infty$ on D^c all Y_n belonging to D^c will be rejected.

2.3   Theorical Results

Two results of convergence can be obtained for the algorithm multistoch. The first proposition deals with the almost sure convergence of the grid point for which the objective function is minimum, if the chosen standard deviation $(\rho_n)_{n \in \mathbb{N}}$ decreases slowly enough. It is interesting to note that this proposition is true for all $T_n > 0$. The second proposition shows a more global convergence of the algorithm when T_n decreases to 0. The proof of these two theorem are given in the appendix.

Proposition 1. *If* $\rho_k = \frac{C}{(\log(k))^\alpha}, k > 1, C > 0$ *with* $\alpha < \frac{1}{2}$ *or* $\alpha = \frac{1}{2}$ *and* $C \geq \frac{\Delta}{\sqrt{2}}$ *then:*

$$\underset{(\Gamma_n^i), i = 1, \ldots, N}{argmin} \quad f(\Gamma_n^i) \quad \underset{n \to +\infty}{\longrightarrow} \quad x^* \quad almost \quad surely,$$

with $\Delta = \underset{x,y \in D}{\max} \|x - y\|$.

The second proposition shows a more global convergence of the algorithm.

Proposition 2. *Under the assumptions of proposition 1 and assuming that* $(T_n)_{n \in \mathbb{N}}$ *decreases to 0 we have:*

$$X_n \quad \overset{\mathbb{P}}{\underset{n \to +\infty}{\longrightarrow}} \quad x^*.$$

2.4   Numerical Results

Here we look at some results obtained on the engine calibration optimization problem of 90 variables. It compares pure random multistart, Random Linkage (see [4]) and the algorithm multistoch. The multistart algorithm is the simpliest algorithm which allies a global and local optimization algorithm, it uses a uniform distribution over D to generate all the starting points for the local optimization algorithm. The random Linkage algorithm generates a new starting points for a local optimization algorithm by taking in accoumpt the preceding starting points. Here multistoch uses a grid of 20 points in D. The computation time for multistoch did not exceeded five minutes.

Table 1. Repartition in percentage of the local optima obtained among five ranges of fuel consumption

Algorithm	[1440,1450]	[1450,1460]	[1460,1470]	[1470,1480]	[1480,1490]
Multistart	29%	30%	25%	15%	1
Random linkage	28%	30%	31%	10%	1
Multistoch	50%	31%	18%	1%	0

Table 1 shows that, for this problem, Multistoch gives better results for a fixed number of local optimization. Moreover if the number of local optimization procedure are limited multistoch obtains a higher number of local optima in range $[1440, 1450]$.

3   Conclusion

The results given by the algorithm are sastisfying for the engine calibration problem and is used in engine calibration procedure. Further work is to compare the algorithm on a academical problem test. Concerning the theorical results of the algorithm the speed of convergence is being explored.

Acknowledgement

Many thanks to Gilles Pagès and Josselin Visconti for their helpful comments.

4   Appendix

This appendix contains the proof of the two proposition, for simplification reasons we note

$$\operatorname*{argmin}_{(\Gamma^i),\, i \in \{1,\ldots,N\}} f(\Gamma^i) = \operatorname{argmin}_{\Gamma^i} f(\Gamma^i)$$

4.1   Proof of Theorem 1

The first step is to show that

$$\mathbb{P}\left(\forall k \geq 0, \quad \operatorname{argmin}_{\Gamma_k^i} f(\Gamma_k^i) \notin V_\varepsilon^f\right) = 0,$$

where $V_\varepsilon^f = f^{-1}\left((f(x^*) - \varepsilon, f(x^*) + \varepsilon)\right)$. We focus on the convergence in probability of the minimal point of the grid.
Let $\varepsilon > 0$ be fixed.
First case: If $\operatorname{argmin}_{\Gamma_0^i} f(\Gamma_0^i) \in V_\varepsilon^f$
As $\operatorname{argmin}_{\Gamma_0^i} f(\Gamma_0^i) \in V_\varepsilon^f$ we have $\forall k \geq 0$,
$\operatorname{argmin}_{\Gamma_k^i} f(\Gamma_k^i) \in V_\varepsilon^f$ almost surely by construction of X_n and of Γ_n, it gives

$$\mathbb{P}\left(\forall k \geq 0, \quad \operatorname{argmin}_{\Gamma_k^i} f(\Gamma_k^i) \notin V_\varepsilon^f\right) = 0.$$

Second case: If $\operatorname{argmin}_{\Gamma_0^i} f(\Gamma_0^i) \notin V_\varepsilon^f$
We can note that

$$\bigcap_{k \geq 0} \left\{\operatorname{argmin}_{\Gamma_k^i} f(\Gamma_k^i) \notin V_\varepsilon^f\right\} \subset \bigcap_{k \geq 1} \{Y_k \notin V_\varepsilon^f\} \cap \left\{\operatorname{argmin}_{\Gamma_0^i} f(\Gamma_0^i) \notin V_\varepsilon^f\right\}$$

$$\bigcap_{k \geq 0} \left\{\operatorname{argmin}_{\Gamma_k^i} f(\Gamma_k^i) \notin V_\varepsilon^f\right\} \subset \bigcap_{k \geq 1} \{Y_k \notin V_\varepsilon^f\}.$$

We need to show

$$\lim_{n \to +\infty} \mathbb{P}\left(\bigcap_{k=1}^{n} \{Y_k \notin V_\varepsilon^f\} \right) = 0.$$

If we note $\mathbb{E}$ the mean function, we have

$$\mathbb{P}\left(\bigcap_{k=1}^{n} \{Y_k \notin V_\varepsilon^f\} \right) = \mathbb{E}\left(\mathbb{E}\left(\mathbb{1}_{\left\{ \bigcap_{k=1}^{n} \{Y_k \notin V_\varepsilon^f\} \right\}} \middle| \mathcal{F}_{n-1} \right) \right)$$

$$\mathbb{P}\left(\bigcap_{k=1}^{n} \{Y_k \notin V_\varepsilon^f\} \right) = \mathbb{E}\left(\mathbb{1}_{\left\{ \bigcap_{k=1}^{n-1} \{Y_k \notin V_\varepsilon^f\} \right\}} \mathbb{E}\left(\mathbb{1}_{\{Y_n \notin V_\varepsilon^f\}} \middle| \mathcal{F}_{n-1} \right) \right). \quad (1)$$

Now

$$\mathbb{P}\left(Y_n \notin V_\varepsilon^f \middle| \mathcal{F}_{n-1} \right) =$$

$$\mathbb{E}\left(\sum_{i=1}^{N} \mathbb{1}_{\{f(\rho_n V_n + \Gamma_{n-1}^i) > \varepsilon + f(x^*)\} \cap \{Z_n = \Gamma_{n-1}^i\} \cap \{f(\Gamma_{n-1}^i) < \min_i f(\Gamma_{n-1}^i) + \beta\}} \middle| \mathcal{F}_{n-1} \right)$$

$$+ \mathbb{E}\left(\sum_{i=1}^{N} \mathbb{1}_{\{f(\rho V_n + \Gamma_{n-1}^i) > \varepsilon + f(x^*)\} \cap \{Z_n = \Gamma_{n-1}^i\} \cap \{f(\Gamma_{n-1}^i) \geq \min_i f(\Gamma_{n-1}^i) + \beta\}} \middle| \mathcal{F}_{n-1} \right)$$

$$\mathbb{P}\left(Y_n \notin V_\varepsilon^f \middle| \mathcal{F}_{n-1} \right) =$$

$$\mathbb{E}\left(\sum_{i=1}^{N} \mathbb{1}_{\{f(\rho_n V_n + \Gamma_{n-1}^i) > \varepsilon + f(x^*)\} \cap \{U_n \in I_{n-1}^i\} \cap \{f(\Gamma_{n-1}^i) < \min_i f(\Gamma_{n-1}^i) + \beta\}} \middle| \mathcal{F}_{n-1} \right)$$

$$+ \mathbb{E}\left(\sum_{i=1}^{N} \mathbb{1}_{\{f(\rho V_n + \Gamma_{n-1}^i) > \varepsilon + f(x^*)\} \cap \{U_n \in I_{n-1}^i\} \cap \{f(\Gamma_{n-1}^i) \geq \min_i f(\Gamma_{n-1}^i) + \beta\}} \middle| \mathcal{F}_{n-1} \right).$$

As $(U_n, V_n)_{n \in \mathbb{N}}$ is independant of $\mathcal{F}_{n-1}$ by construction we have

$$\mathbb{P}\left(Y_n \notin V_\varepsilon^f \middle| \mathcal{F}_{n-1} \right) = \sum_{i=1}^{N} \mathbb{P}\left(U_n \in I_{n-1}^i \right)$$

$$\mathbb{P}\left(\{f(\rho_n V_n + \Gamma_{n-1}^i) > \varepsilon + f(x^*)\} \cap \{f(\Gamma_{n-1}^i) < \min_i f(\Gamma_{n-1}^i) + \beta\} \middle| \mathcal{F}_{n-1} \right)$$

$$+ \sum_{i=1}^{N} \mathbb{P}\left(U_n \in I_{n-1}^i \right)$$

$$\mathbb{P}\left(\{f(\rho V_n + \Gamma_{n-1}^i) > \varepsilon + f(x^*)\} \cap \{f(\Gamma_{n-1}^i) \geq \min_i f(\Gamma_{n-1}^i) + \beta\} \middle| \mathcal{F}_{n-1} \right)$$

but $\{f(\Gamma_{n-1}^i) < \min_i f(\Gamma_{n-1}^i) + \beta\}$ belong to $\mathcal{F}_{n-1}$ we get

$$\mathbb{P}\left(Y_n \notin V_\varepsilon^f | \mathcal{F}_{n-1}\right) = \sum_{i=1}^{N} \mathbb{P}\left(U_n \in I_{n-1}^i\right) \mathrm{1\!I}_{\{f(\Gamma_{n-1}^i) < \min_i f(\Gamma_{n-1}^i) + \beta\}}$$

$$\mathbb{P}\left(f(\rho_n V_n + \Gamma_{n-1}^i) > \varepsilon + f(x^*) | \mathcal{F}_{n-1}\right)$$

$$+ \sum_{i=1}^{N} \mathbb{P}\left(U_n \in I_{n-1}^i\right) \mathrm{1\!I}_{\{f(\Gamma_{n-1}^i) \geq \min_i f(\Gamma_{n-1}^i) + \beta\}}$$

$$\mathbb{P}\left(f(\rho V_n + \Gamma_{n-1}^i) > \varepsilon + f(x^*) | \mathcal{F}_{n-1}\right).$$

If we define

$$\theta = \frac{\exp\left(\frac{-\Delta^2}{2\rho^2}\right)}{(2\pi\rho^2)^{\frac{d}{2}}}$$

$$\theta_n = \frac{\exp\left(\frac{-\Delta^2}{2\rho_n^2}\right)}{(2\pi\rho_n^2)^{\frac{d}{2}}}$$

we get

$$\mathbb{P}\left(f(\rho_n V_n + \Gamma_{n-1}^i) \leq \varepsilon + f(x^*) | \mathcal{F}_{n-1}\right) \geq \min(\theta_n, \theta)\lambda_d(V_\varepsilon^f)$$

where λ_d is the lebesgue measure on $\mathbb{R}^d$; We get

$$\mathbb{P}\left(f(\rho V_n + \Gamma_{n-1}^i) \leq \varepsilon + f(x^*) | \mathcal{F}_{n-1}\right) \geq \min(\theta_n, \theta)\lambda_d(V_\varepsilon^f).$$

Which gives

$$\mathbb{P}\left(Y_n \notin V_\varepsilon^f | \mathcal{F}_{n-1}\right) \leq 1 - \sum_{i=1}^{N} \mathbb{P}(U_n \in I_{n-1}^i) \min(\theta_n, \theta)\lambda_d(V_\varepsilon^f)$$

$$\mathbb{P}\left(Y_n \notin V_\varepsilon^f | \mathcal{F}_{n-1}\right) \leq 1 - \min(\theta_n, \theta)\lambda_d(V_\varepsilon^f).$$

By noticing that the function $x \longmapsto \frac{1}{(2\pi x^2)^{\frac{d}{2}}} \exp\left(\frac{-\Delta^2}{2x^2}\right)$ is increasing for $x \in [0, \frac{\Delta}{\sqrt{d}}]$. the sequence θ_k decreases, for all k large enough. Consequently, there exists $M_0 > N_0$ such as

$$\min(\theta_n, \theta) = \theta_n, \quad \forall k > M_0,$$

we get

$$\mathbb{P}\left(Y_n \notin V_\varepsilon^f | \mathcal{F}_{n-1}\right) \leq 1 - \lambda_d(V_\varepsilon^f)\theta_n, \quad \forall n > M_0.$$

Therefore we can write using equation (1) that there exists a positive constant C_{M_0} such as

$$\mathbb{P}\left(\bigcap_{k=1}^{n}\{Y_k \notin V_\varepsilon^f\}\right) \leq C_{M_0} \prod_{k=M_0+1}^{n} (1 - \lambda_d(V_\varepsilon^f)\theta_k)$$

$$\mathbb{P}\left(\bigcap_{k=1}^{n}\{Y_k \notin V_\varepsilon^f\}\right) \le C_{M_0}\exp\left(-\sum_{k=M_0+1}^{n}\lambda_d(V_\varepsilon^f)\theta_k\right). \qquad (2)$$

It can be showned that if $\rho_k = \frac{C}{(\log(k))^\alpha}$, $k > 1$ and $C > 0$ that

- if $\alpha < \frac{1}{2}$ then $\sum_{k>1}\theta_k = +\infty$,
- if $\alpha > \frac{1}{2}$ then $\sum_{k>1}\theta_k < +\infty$,
- if $\alpha = \frac{1}{2}$ and
    - if $C < \frac{A}{\sqrt{2}}$ then $\sum_{k>1}\theta_k < +\infty$,
    - if $C \ge \frac{A}{\sqrt{2}}$ then $\sum_{k>1}\theta_k = +\infty$.

Which gives

$$\forall \varepsilon > 0, \quad \mathbb{P}\left(\bigcap_{k=1}^{n}\{Y_k \notin V_\varepsilon^f\}\right) \underset{n \to +\infty}{\longrightarrow} 0$$

and $\bigcap_{k=1}^{n}\{Y_k \notin V_\varepsilon^f\} \supset \left\{\text{argmin}_{\Gamma_k^i} f(\Gamma_k^i) \notin V_\varepsilon^f\right\}$, consequently

$$\mathbb{P}\left(\forall k \ge 0, \quad \text{argmin}_{\Gamma_k^i} f(\Gamma_k^i) \notin V_\varepsilon^f\right) = 0.$$

And finally

$$\text{argmin}_{\Gamma_k^i} f(\Gamma_k^i) \underset{k \to +\infty}{\overset{\mathbb{P}}{\longrightarrow}} x^*.$$

Almost sure convergence is obtained by noticing that $\min_{i \in \{1,\dots,N\}} f(\Gamma_n^i)$ is a decreasing sequence superior to $f(x^*)$. Which means that the sequence converges almost surely towards a limit that can only be x^*.

4.2 Proof of Theorem 2

We define $\varepsilon > 0$ and $\delta > 0$. The first step of the proof will show the existence of M such that
$$\forall n > M \quad \mathbb{P}(Z_n \notin V_\varepsilon^f) < \delta.$$

We note

$$a_n = \mathbb{P}\left(\{Z_n \notin V_\varepsilon^f\} \cap \left\{\text{argmin}_{\Gamma_{n-1}^i} f(\Gamma_{n-1}^i) \in V_{\frac{\varepsilon}{2}}^f\right\}\right).$$

We have

$$a_n = \mathbb{P}\left(\{f(Z_n) \geq f(x^*) + \varepsilon\} \cap \left\{f(\Gamma^i_{n-1}) \in V^f_{\frac{\varepsilon}{2}}\right\}\right)$$

$a_n =$
$$\mathbb{P}\left(\left\{\exp(\frac{-(f(x^*)+\varepsilon)}{T_n}) \geq \exp(\frac{-f(Z_n)}{T_n})\right\} \cap \left\{\text{argmin}_{\Gamma^i_{n-1}} f(\Gamma^i_{n-1}) \in V^f_{\frac{\varepsilon}{2}}\right\}\right)$$

$$\mathbb{E}\left(\sum_{i=1}^N \mathbb{1}_{\left\{\exp(\frac{-(f(x^*)+\varepsilon)}{T_n}) \geq \exp(\frac{-f(\Gamma^i_{n-1})}{T_n})\right\}} \cap \{Z_n = \Gamma^i_{n-1}\} \cap \left\{\text{argmin}_{\Gamma^i_{n-1}} f(\Gamma^i_{n-1}) \in V^f_{\frac{\varepsilon}{2}}\right\}\right)$$

$$\mathbb{E}\left(\sum_{i=1}^N \mathbb{1}_{\left\{\exp(\frac{-(f(x^*)+\varepsilon)}{T_n}) \geq \exp(\frac{-f(\Gamma^i_{n-1})}{T_n})\right\}} \cap \{U_n \in I^i_{n-1}\} \cap \left\{\text{argmin}_{\Gamma^i_{n-1}} f(\Gamma^i_{n-1}) \in V^f_{\frac{\varepsilon}{2}}\right\}\right)$$

noticing that U_n and $\mathcal{F}_{n-1}$ are independant and $\left\{f(\Gamma^i_{n-1}) \in V^f_{\frac{\varepsilon}{2}}\right\} \in \mathcal{F}_{n-1}$, if we put $A = \left\{\text{argmin}_{\Gamma^i_{n-1}} f(\Gamma^i_{n-1}) \in V^f_{\frac{\varepsilon}{2}}\right\}$ we get

$a_n \leq$
$$\mathbb{E}\left(\sum_{i=1}^N \frac{\exp(-f(\Gamma^i_{n-1})/T_n)}{\sum_{j=1}^N \exp(-f(\Gamma^j_{n-1})/T_n)} \mathbb{1}_{\{\exp(-(f(x^*)+\varepsilon)/T_n) \geq \exp(-f(\Gamma^i_{n-1})/T_n)\} \cap \{A\}}\right)$$

$$a_n \leq N \exp\left(\frac{-(f(x^*)+\varepsilon)}{T_n}\right) \mathbb{E}\left(\frac{1}{\sum_{j=1}^N \exp(\frac{-f(\Gamma^j_{n-1})}{T_n})} \mathbb{1}_A\right)$$

$$a_n \leq N \exp\left(\frac{-\varepsilon}{2T_n}\right). \tag{3}$$

As T_n is decreasing there exists $n_0 > 0$ such that for all $n > n_0$

$$a_n < \frac{\delta}{2}.$$

We now focus on $\mathbb{P}(Z_n \notin V^f_\varepsilon)$. We can note that

$$\mathbb{P}\left(Z_n \notin V^f_\varepsilon\right) = \mathbb{P}\left(\{Z_n \notin V^f_\varepsilon\} \cap \left\{\text{argmin}_{\Gamma^i_{n-1}} f(\Gamma^i_{n-1}) \in V^f_{\frac{\varepsilon}{2}}\right\}\right) +$$
$$\mathbb{P}\left(\{Z_n \notin V^f_\varepsilon\} \cap \left\{\text{argmin}_{\Gamma^i_{n-1}} f(\Gamma^i_{n-1}) \notin V^f_{\frac{\varepsilon}{2}}\right\}\right)$$

$$\mathbb{P}\left(Z_n \notin V^f_\varepsilon\right) \leq \mathbb{P}\left(\{Z_n \notin V^f_\varepsilon\} \cap \left\{\text{argmin}_{\Gamma^i_{n-1}} f(\Gamma^i_{n-1}) \in V^f_{\frac{\varepsilon}{2}}\right\}\right) +$$
$$\mathbb{P}\left(\text{argmin}_{\Gamma^i_{n-1}} f(\Gamma^i_{n-1}) \notin V^f_{\frac{\varepsilon}{2}}\right).$$

Now by the previous theorem there exists $n_1 > 0$ such that all $n \geq n_1$

$$\mathbb{P}\left(\text{argmin}_{\Gamma^i_{n-1}} f(\Gamma^i_{n-1}) \notin V^f_{\frac{\varepsilon}{2}}\right) < \frac{\delta}{2}$$ and for all $n > M = \max(n_0, n_1)$ we have

$$\mathbb{P}(Z_n \notin V^f_\varepsilon) \leq \delta.$$

Consequently we obtain

$$\exists M \quad \forall n > M \quad \mathbb{P}(Z_n \notin V^f_\varepsilon) \leq \delta$$

By noticing that $f(X_n) < f(Z_n)$ that is

$$\{X_n \notin V^f_\varepsilon\} \subset \{Z_n \notin V^f_\varepsilon\}$$

we can conclude. □

References

[1] Hafner, M., Isermann, R.: The use of stationary and dynamic emission models for an improved engine cycles. International Workshop on Modelling Emissions and Control in Automotive Engines, Salerno, Italy, 09/2000.

[2] Herskovits, J. N.: A feasible Directions Interior point Technique for Nonlinear Optimization. Journal of Optimization Theory and Applications, vol. 99, n:1, pp.121-146, October, 1998.

[3] Locatelli, M.: Convergence of a simulated annealing for continuous global optimization. Journal of Global Optimization, 18, 219-233, 2000.

[4] Locatelli, M., Schoen, F.: Random Linkage: a family of acceptance/rejection algorithms for global optimization. Mathematical Programming, 85, 2, 379-396, 1999.

Numerical Behavior of a Stabilized SQP Method for Degenerate NLP Problems

El–Sayed M. E. Mostafa[1,*,**],
Luís N. Vicente[2,*,***], and Stephen J. Wright[3,†]

[1] Centro de Matemática
Universidade de Coimbra, 3001-454 Coimbra, Portugal.
emostafa@mat.uc.pt.
[2] Departamento de Matemática
Universidade de Coimbra
3001-454 Coimbra, Portugal
lnv@mat.uc.pt.
[3] Department of Computer Sciences
University of Wisconsin
Madison, WI 53706, USA
swright@cs.wisc.edu

Abstract. In this paper we discuss the application of the stabilized SQP method with constraint identification (sSQPa) recently proposed by S. J. Wright [12] for nonlinear programming problems at which strict complementarity and/or linear independence of the gradients of the active constraints may fail to hold at the solution. We have collected a number of degenerate problems from different sources. Our numerical experiments have shown that the sSQPa is efficient and robust even without the incorporation of a classical globalization technique. One of our goals is therefore to handle NLPs that arise as subproblems in global optimization where degeneracy and infeasibility are important issues. We also discuss and present our work along this direction.

Keywords: nonlinear programming, successive quadratic programming, degeneracy, identification of active constraints, infeasibility

[*] Support for the first two authors was provided by Centro de Matemática da Universidade de Coimbra, by FCT under grant POCTI/35059/MAT/2000, and by the European Union under grant IST-2000-26063.

[**] On leave from the Department of Mathematics, Faculty of Science, Alexandria University, Alexandria, Egypt

[***] Correspondent author

[†] Support for this author was provided by the Mathematical, Information, and Computational Sciences Division subprogram of the Office of Advanced Scientific Computing, U.S. Department of Energy under contract W-31-109-Eng-38, and the National Science Foundation under grants CDA-9726385 and ACI-0082065.

C. Bliek et al. (Eds.): COCOS 2002, LNCS 2861, pp. 123–141, 2003.
© Springer-Verlag Berlin Heidelberg 2003

1 Introduction

We consider the general nonlinear programming (NLP) problem written in the form

$$\min_{z} \ \phi(z) \qquad \text{subject to} \quad g_I(z) \leq 0, \quad g_E(z) = 0, \tag{1}$$

where $\phi : \mathbb{R}^n \to \mathbb{R}$, $g_I : \mathbb{R}^n \to \mathbb{R}^{m_I}$, $g_E : \mathbb{R}^n \to \mathbb{R}^{m_E}$ ($m_E = m - m_I$; m the total number of constraints) are assumed to be twice continuously differentiable functions. Let us write the two subsets of indices for the inequality and equality constraints, respectively, as

$$I = \{i = 1, \ldots, m_I\}, \qquad E = \{i = m_I + 1, \ldots, m\}.$$

Throughout this article the subsets I and E refer to the elements in the subsets of inequality and equality constraints. The Lagrangian function associated with this problem is

$$\mathcal{L}(z, \lambda) = \phi(z) + \lambda_I^T g_I(z) + \lambda_E^T g_E(z),$$

where $\lambda = (\lambda_I, \lambda_E) \in \mathbb{R}^m$ are the Lagrange multipliers associated with the inequality and equality constraints in (1). For simplicity we write the vector of the multipliers as (λ_I, λ_E), while the accurate form would be $(\lambda_I^T, \lambda_E^T)^T$. The default norm in this paper is the ℓ_2.

2 Assumptions, Notations, and Basic Results

The local convergence theory of the stabilized SQP method with constraint identification is based on the following assumptions [12]:

Assumption 1. *Let ϕ and g be twice Lipschitz continuously differentiable in a neighborhood of a point z^*. Let the Mangasarian-Fromovitz constraint qualification, the first-order necessary optimality conditions, and a form of second-order sufficient optimality conditions hold at z^*.*

Note that there is no assumption made about the linear independence of the gradients of the active constraints. Since the vector of optimal Lagrange multipliers is not unique if the gradients of the active constraints are linearly dependent, we need to consider the set of optimal Lagrange multipliers, denoted by $\mathcal{S}_\lambda$:

$$\mathcal{S}_\lambda = \{\lambda \mid \nabla_z \mathcal{L}(z^*, \lambda) = 0, \ \lambda_I^T g_I(z^*) = 0, \ \lambda_I \geq 0\}.$$

The optimal primal-dual set consists of the pairs (z, λ) in

$$\mathcal{S} = \{z^*\} \times \mathcal{S}_\lambda.$$

We also remark that there is no assumption about strict complementarity between z^* and the elements in $\mathcal{S}_\lambda$.

We need now several definitions to describe the stabilized SQP method with constraint identification. The set of active inequality constraints at z^* is defined as

$$\mathcal{B} = \{i = 1, \ldots, m_I \mid g_{I_i}(z^*) = 0\}.$$

For any optimal multipliers $\lambda^* \in \mathcal{S}_\lambda$ we define the set

$$\mathcal{B}_+(\lambda^*) = \{i \in \mathcal{B} \mid \lambda^*_{I_i} > 0\}.$$

The set of strong active constraints and the set of weak active constraints are defined as:

$$\mathcal{B}_+ = \bigcup_{\lambda^* \in \mathcal{S}_\lambda} \mathcal{B}_+(\lambda^*), \qquad \mathcal{B}_0 = \mathcal{B} \backslash \mathcal{B}_+.$$

The distance of a pair (z, λ) to the optimal primal-dual set $\mathcal{S}$ is denoted by $\delta(z, \lambda)$:

$$\delta(z, \lambda) = \text{dist}((z, \lambda), \mathcal{S}),$$

where

$$\text{dist}((z, \lambda), \mathcal{S}) = \inf_{(z^*, \lambda^*) \in \mathcal{S}} \|(z^*, \lambda^*) - (z, \lambda)\|.$$

3   Stabilized SQP with Constraint Identification

SQP methods have shown to be quite successful in solving NLP problems. For degenerate NLP problems, where at the solution the linear independence of the gradients of the active constraints and/or the strict complementarity condition may not hold, Wright [12] has designed a stabilized SQP method, *algorithm sSQPa*, to handle such type of problems. For that purpose, a constraint identification procedure, *procedure ID0*, has been developed to identify the active inequality constraints and, furthermore, to classify them as strong or weak active constraints. The method also considers the solution of an LP subproblem to provide an *interior* multipliers estimate and a stabilization of the traditional QP subproblem. Both the constraint identification procedure and the interior multipliers estimate are called when no sufficient reduction is obtained by the solution of the stabilized QP subproblem. Wright [12] proved a superlinear rate of local convergence for algorithm sSQPa with procedure ID0.

In the next subsections, we give a complete description of the overall algorithm, extending the presentation of Wright [12] to the general case of inequality and equality constraints.

3.1   SQP Algorithm

Standard SQP methods for the NLP problem (1) typically solve a sequence of QP subproblems of the following form

$$\min_{\Delta z} \ \nabla \phi(z)^T \Delta z + \frac{1}{2} \Delta z^T \nabla_{zz} \mathcal{L}(z, \lambda) \Delta z$$

$$\text{subject to} \ g_I(z) + \nabla g_I(z)^T \Delta z \le 0, \qquad g_E(z) + \nabla g_E(z)^T \Delta z = 0, \qquad (2)$$

where (z, λ) is the current iterate. The stabilized SQP method [12] considers instead the following minimax subproblem

$$\min_{\Delta z} \max_{\lambda_I^+ \geq 0, \lambda_E^+} \nabla\phi(z)^T \Delta z + \frac{1}{2}\Delta z^T \nabla_{zz}\mathcal{L}(z, \lambda)\Delta z + (\lambda_I^+)^T[g_I(z) + \nabla g_I(z)^T \Delta z]$$

$$+ (\lambda_E^+)^T[g_E(z) + \nabla g_E(z)^T \Delta z] - \frac{\mu}{2}\|\lambda^+ - \lambda\|^2,$$

where $\mu \geq 0$ is a given parameter, and the solution $\lambda^+ = (\lambda_I^+, \lambda_E^+)$ provides a new update for the Lagrange multipliers associated with the inequality and equality constraints. This minimax subproblem is in turn equivalent to the following QP subproblem

$$\min_{(\Delta z, \lambda^+)} \nabla\phi(z)^T \Delta z + \frac{1}{2}\Delta z^T \nabla_{zz}\mathcal{L}(z, \lambda)\Delta z + \frac{\mu}{2}\|\lambda^+\|^2 \tag{3}$$

$$\text{subject to } g_E(z) + \nabla g_E(z)^T \Delta z - \mu(\lambda_E^+ - \lambda_E) = 0,$$
$$g_I(z) + \nabla g_I(z)^T \Delta z - \mu(\lambda_I^+ - \lambda_I) \leq 0,$$

with $\lambda^+ = (\lambda_I^+, \lambda_E^+)$; see [6], [12]. One can easily see that the QP subproblem (3) is posed in both z and λ^+ and has therefore m variables more than the traditional QP subproblem (2).

The stabilizing parameter μ introduced in the above subproblems is chosen as $\mu = \eta(z, \lambda)^\sigma$ with $\sigma \in (0, 1)$, where $\eta(z, \lambda)$ is the size of the residual of the first-order necessary conditions given by

$$\eta(z, \lambda) = \left\| \begin{bmatrix} \nabla_z\mathcal{L}(z, \lambda) \\ \min(\lambda_I, -g_I(z)) \\ g_E(z) \end{bmatrix} \right\|,$$

with the min operator applied component-wise. In fact, $\eta(z, \lambda)$ represents a practical way of measuring the distance to the primal-dual set $\mathcal{S}$; see e.g., [12, Theorem 2].

The quantity $\eta(z, \lambda)$ also provides an estimate for the set of active constraints $\mathcal{B}$:

$$\mathcal{A}(z, \lambda) = \{i = 1, 2, \ldots, m_I \mid g_{I_i}(z) \geq -\eta(z, \lambda)^\tau\}, \quad \tau \in (0, 1), \tag{4}$$

see [12]. It is clear that when (z, λ) approaches a primal-dual solution, then the distance $\delta(z, \lambda)$ decreases and the interval of feasibility measured by the lower bound $-\eta(z, \lambda)^\tau$ reduces too, improving the quality of the estimation provided by $\mathcal{A}(z, \lambda)$. In addition, the estimated set $\mathcal{A}(z, \lambda)$ is partitioned into a subset $\mathcal{A}_+$ of estimated strong active constraints and a subset $\mathcal{A}_0$ of estimated weak active constraints. Depending on the decrease on $\eta(z, \lambda)$ provided by the solution of the QP subproblem (3), an LP subproblem is solved in order to maximize the multipliers corresponding to the inequality constraints in the subset $\mathcal{A}_+$, keeping the remaining multipliers corresponding to inequality constraints at zero. The identification procedure and the interior multipliers estimation will be introduced in subsections 3.2 and 3.3, respectively. Now, we restate the algorithm sSQPa [12].

Algorithm sSQPa

Choose parameters $\tau, \sigma \in (0,1)$, a tolerance $\mathtt{tol} > 0$, and an initial starting point (z^0, λ^0) with $\lambda_I^0 \geq 0$. Compute $\mathcal{A}(z^0, \lambda^0)$ using (4), call procedure ID0 to compute the subsets $\mathcal{A}_+$ and $\mathcal{A}_0$, and solve the LP subproblem (6) to obtain $\hat{\lambda}^0$. Set $k \leftarrow 0$ and $\lambda^0 \leftarrow \hat{\lambda}^0$.

While $\eta(z^k, \lambda^k) > \mathtt{tol}$ do

    Solve (3) for $(\Delta z, \lambda^+)$, and set $\mu^k = \eta(z^k, \lambda^k)^\sigma$.

    If $\eta(z^k + \Delta z, \lambda^+) \leq \left(\eta(z^k, \lambda^k)\right)^{1+\sigma/2}$

        set $(z^{k+1}, \lambda^{k+1}) \leftarrow (z^k + \Delta z, \lambda^+)$; set $k \leftarrow k + 1$;

    else

        compute $\mathcal{A}(z^k, \lambda^k)$, and then apply ID0 to obtain $\mathcal{A}_+$ and $\mathcal{A}_0$;

        solve the LP subproblem (6) to obtain $\hat{\lambda}^k$, and set $\lambda^k \leftarrow \hat{\lambda}^k$;

    end(if)

end(while)

For each iterate (z^k, λ^k) one solves the QP subproblem (3). If the computed step $(\Delta z, \lambda^+)$ yields a sufficient decrease in $\eta(z, \lambda)$, then $(\Delta z, \lambda^+)$ is accepted, otherwise $(\Delta z, \lambda^+)$ is rejected and the sSQPa algorithm switches to its $\mathtt{else}$ condition. In such a case, the set $\mathcal{A}(z, \lambda)$ is updated and the procedure ID0 is called to partition the set $\mathcal{A}(z, \lambda)$ into the subsets $\mathcal{A}_+$ and $\mathcal{A}_0$. A new multipliers estimate $\hat{\lambda}^k$ is computed by the solution of an LP subproblem.

The next result shows that the rate of local convergence of algorithm sSQPa is superlinear for degenerate problems [12, Theorem 7]. It is also shown that when (z^0, λ^0) is close to the optimal set $\mathcal{S}$, the initial call of procedure ID0 is the only one that is needed. The numerical experiments presented in this paper confirm these statements.

Theorem 1. *Suppose that assumption 1 holds. Then there exists a constant $\bar{\delta} > 0$ such that for any (z^0, λ^0) with $\delta(z^0, \lambda^0) \leq \bar{\delta}$, the $\mathtt{if}$ condition in algorithm sSQPa is always satisfied and the sequence $\{\delta(z^k, \lambda^k)\}$ converges superlinearly to zero with q-order $1 + \sigma$.*

3.2   Constraint Identification

The set $\mathcal{A}(z, \lambda)$ defined in (4) has been used to estimate the active inequality constraints in a neighborhood of a solution, see [2], [12]. In this estimation all inequality constraints with function values greater than or equal to $-\eta(z, \lambda)^\tau$ are considered in $\mathcal{A}(z, \lambda)$. Under the standing assumptions it can be shown [12, Theorem 3] that in a sufficiently small neighborhood of the solution the set $\mathcal{A}(z, \lambda)$ successfully estimates the active set $\mathcal{B}$.

Lemma 1. *Let assumption 1 holds. Then, there exists $\delta_1 > 0$ such that for all (z, λ) with $\delta(z, \lambda) \leq \delta_1$, it holds $\mathcal{A}(z, \lambda) = \mathcal{B}$.*

As we have said before, it is also desirable to partition the set $\mathcal{A}(z, \lambda)$ in two sets: one corresponding to constraints that are candidate to be strong and the other containing the constraints that are candidates to be weak. To achieve this purpose it is convenient to solve the following LP subproblem [12] for a given subset $\hat{\mathcal{A}} \subset \mathcal{A}(z, \lambda)$ containing the candidates for weak active constraints.

$$\max_{\tilde{\lambda}_I, \tilde{\lambda}_E} \sum_{i \in \hat{\mathcal{A}}} \tilde{\lambda}_i$$

$$\text{subject to} \quad \left\| \nabla \phi(z) + \sum_{i \in \mathcal{A}(z,\lambda)} \tilde{\lambda}_{I_i} \nabla g_{I_i}(z) + \sum_{i \in E} \tilde{\lambda}_{E_i} \nabla g_{E_i}(z) \right\|_\infty \leq \chi(z, \lambda, \tau),$$

$$\tilde{\lambda}_{I_i} \geq 0 \quad \text{for all} \quad i \in \mathcal{A}(z, \lambda), \quad \tilde{\lambda}_{I_i} = 0 \quad \text{for all} \quad i \in I \backslash \mathcal{A}(z, \lambda),$$

where $\chi(z, \lambda, \tau)$ is given by

$$\chi(z, \lambda, \tau) =$$

$$\max \left(\eta(z, \lambda)^\tau, \left\| \nabla \phi(z) + \sum_{i \in \mathcal{A}(z,\lambda)} \lambda_{I_i} \nabla g_{I_i}(z) + \sum_{i \in E} \lambda_{E_i} \nabla g_{E_i}(z) \right\|_\infty \right). \quad (5)$$

The multipliers $\hat{\lambda}_E$ corresponding to the equality constraints have no sign restriction in this LP subproblem.

In the following lines we restate the constraint identification procedure ID0 proposed in [12] based on the solution of LP subproblems of this type. The output of procedure ID0 is a partition of $\mathcal{A}(z, \lambda)$ into two sets $\mathcal{A}_+$ and $\mathcal{A}_0$: $\mathcal{A}_+$ contains the candidates for strong active constraints and $\mathcal{A}_0$ the candidates for weak active constraints.

Procedure ID0
Given τ, $\hat{\tau}$ with $0 < \hat{\tau} < \tau < 1$ and a point (z, λ), compute $\chi(z, \lambda, \tau)$ from (5), $\xi(z, \lambda, \tau, \hat{\tau}) = \max \left(\eta(z, \lambda)^{\hat{\tau}}, \chi(z, \lambda, \tau) \right)$, and $\mathcal{A}(z, \lambda)$ from (4). Define $\hat{\mathcal{A}}_{\text{init}} = \mathcal{A}(z, \lambda) \backslash \{i \mid \lambda_{I_i} \geq \xi(z, \lambda, \tau, \hat{\tau})\}$ and set $\hat{\mathcal{A}} \leftarrow \hat{\mathcal{A}}_{\text{init}}$.
Repeat

> If $\hat{\mathcal{A}} = \emptyset$, stop with $\mathcal{A}_0 = \emptyset$ and $\mathcal{A}_+ = \mathcal{A}(z, \lambda)$.
> Solve the LP (5) for $\tilde{\lambda}$ and set $\mathcal{C} = \{i \in \hat{\mathcal{A}} \mid \tilde{\lambda}_{I_i} \geq \xi(z, \lambda, \tau, \hat{\tau})\}$.
> If $\mathcal{C} = \emptyset$
> > stop with $\mathcal{A}_0 = \hat{\mathcal{A}}$ and $\mathcal{A}_+ = \mathcal{A}(z, \lambda) \backslash \hat{\mathcal{A}}$;
> else
> > set $\hat{\mathcal{A}} \leftarrow \hat{\mathcal{A}} \backslash \mathcal{C}$;
> end(if)

end(repeat)

One can see that procedure ID0 will not be exited unless the set $\mathcal{C}$ is empty. The idea is to start with a superset of $\mathcal{A}_0$ given by $\hat{\mathcal{A}} = \hat{\mathcal{A}}_{\text{init}}$ and to remove iteratively from $\hat{\mathcal{A}}$ the constraints, stored in $\mathcal{C}$, that have been estimated to be strong by the LP subproblem (5).

It is shown in [12, Theorem 4] that the two subsets, $\mathcal{A}_+$ and $\mathcal{A}_0$, produced by procedure ID0 successfully estimate $\mathcal{B}_+$ and $\mathcal{B}_0$ in the vicinity of z^*.

Lemma 2. *Let assumption 1 holds. Then, there exists $\delta_2 > 0$ such that whenever $\delta(z, \lambda) \leq \delta_2$, procedure ID0 terminates with $\mathcal{A}_+ = \mathcal{B}_+$ and $\mathcal{A}_0 = \mathcal{B}_0$.*

3.3   Interior Multipliers Estimate

After the application of the constraint identification procedure ID0, the partition of $\mathcal{A}(z, \lambda)$ into the two subsets $\mathcal{A}_+$ and $\mathcal{A}_0$ is available. It is therefore possible to try to make the multipliers corresponding to the estimated strong active constraints in $\mathcal{A}_+$ as far from zero as possible. This is particularly desirable when solving NLP problems arising as subproblems in global optimization. Such interior multipliers estimate can be obtained by solving an LP subproblem of the following form (see [12]), adapted here to include the equality constraints:

$$\max_{\hat{t}, \hat{\lambda}_I, \hat{\lambda}_E} \quad \hat{t}$$

$$\text{subject to} \quad \hat{t} \leq \hat{\lambda}_{I_i} \quad \text{for all} \quad i \in \mathcal{A}_+,$$

$$-\mu \, \mathbf{e} \leq \nabla \phi(z) + \sum_{i \in \mathcal{A}_+} \hat{\lambda}_{I_i} \nabla g_{I_i}(z) + \sum_{i \in E} \hat{\lambda}_{E_i} \nabla g_{E_i}(z) \leq \mu \, \mathbf{e},$$

$$\hat{\lambda}_{I_i} \geq 0 \quad \text{for all} \quad i \in \mathcal{A}_+, \qquad \hat{\lambda}_{I_i} = 0 \quad \text{for all} \quad i \in I \backslash \mathcal{A}_+, \quad (6)$$

where $\mathbf{e}$ is a vector whose entries are all ones and the variables $\hat{\lambda}_E$ are unrestricted in sign.

Under the standing assumptions, it is shown in [12, Theorem 5] that the LP subproblem (6) is feasible and bounded in a sufficiently small neighborhood of the solution. Furthermore, the distance $\delta(z, \hat{\lambda})$ is bounded above by a multiple of $\delta(z, \lambda)^\tau$.

Lemma 3. *Let assumption 1 holds. Then, there exists $\delta_3 > 0$ such that for all (z, λ) with $\delta(z, \lambda) \leq \delta_3$ the LP subproblem (6) is feasible, bounded, and its optimal objective is greater than or equal to*

$$\epsilon_{\lambda^*} = \max_{\lambda^* \in \mathcal{S}_\lambda} \min_{i \in \mathcal{B}_+} \lambda_i^*.$$

Furthermore, there exists $\beta > 0$ such that $\delta(z, \hat{\lambda}) \leq \beta \delta(z, \lambda)^\tau$.

If there exists linear dependency of the gradients of the active constraints, then the vector of optimal Lagrange multipliers is not unique, and one can think of computing the multipliers with the largest possible size. This goal would be particularly relevant when we consider NLP problems arising as subproblems of an enumeration scheme applied to a global optimization problem. With this purpose in mind we have studied a few strategies. The one that seems most relevant consists of solving a second LP subproblem once (6) has been solved. The idea is to maximize the size of the multipliers in $\mathcal{A}_+$ while keeping the lower bound $\hat{t} > 0$ in the infinity norm that has been achieved by solving (6). So, after solving (6), one could solve the following LP subproblem:

$$\max_{\hat{\lambda}_I, \hat{\lambda}_E} \sum_{i \in \mathcal{A}_+} \hat{\lambda}_{I_i}$$

subject to $\quad \hat{t} \leq \hat{\lambda}_{I_i} \quad$ for all $\quad i \in \mathcal{A}_+,$

$$-\mu\, \mathbf{e} \leq \nabla\phi(z) + \sum_{i \in \mathcal{A}_+} \hat{\lambda}_{I_i} \nabla g_{I_i}(z) + \sum_{i \in E} \hat{\lambda}_{E_i} \nabla g_{E_i}(z) \leq \mu\, \mathbf{e},$$

$$\hat{\lambda}_{I_i} \geq 0 \quad \text{for all} \quad i \in \mathcal{A}_+, \qquad \hat{\lambda}_{I_i} = 0 \quad \text{for all} \quad i \in I \backslash \mathcal{A}_+.$$

The numerical experiments have shown, however, that there is not too much gain in solving this second LP subproblem. In fact, the LP subproblem (6) has produced in most instances multipliers whose size was quite close to the largest one.

4   Numerical Experiments

We have developed a Matlab implementation of the stabilized SQP method with constraint identification (algorithm sSQPa) and tested it for a variety of degenerate problems. We used Matlab to solve the LPs and QPs that are needed by the sSQPa method.

We divide the numerical results into three major subsections. In subsection 4.1 we are concerned with the speed of local convergence of the method as well as with its global behavior without any globalization strategy. In subsection 3.2 we describe the numerical performance of procedure ID0 within algorithm sSQPa. Subsection 4.3 describes the use of the sSQPa method to find feasible points for feasible degenerate problems and least infeasible points for infeasible degenerate problems.

4.1   Problems with Objective Function

In this subsection, we consider 12 degenerate NLP test problems. For every problem we tested three different starting points with increasing distance to z_*, and for each case we plot $\log_{10} \|z - z_*\|$ vs iteration number. The performance of sSQPa for each problem is also shown by plotting $\log_{10} \eta(z, \lambda)$ vs iteration number for the farthest away starting point. When possible we compare the performance of sSQPa with other solvers for NLP.

The numerical results are obtained without any globalization strategy. The stopping criterion is $\eta(z, \lambda) \leq 10^{-8}$. We have computed the initial Lagrange

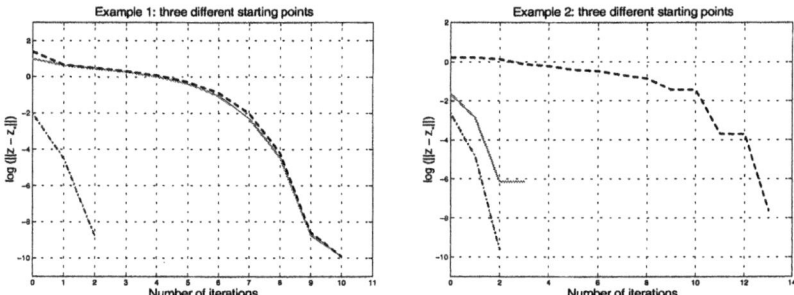

Fig. 1. Convergence from three different starting points for problems 1 and 2

multipliers λ^0 by solving the least-squares problem

$$\min_{\lambda_{\mathcal{A}_{ls}},\lambda_E} \left\| \nabla\phi(z^0) + \nabla g(z^0)_{\mathcal{A}_{ls}}^T \lambda_{\mathcal{A}_{ls}} + \nabla g(z^0)_E^T \lambda_E \right\|^2 \quad \text{s.t.} \quad \lambda_{\mathcal{A}_{ls}} \geq 0, \qquad (7)$$

where $\mathcal{A}_{ls}$ is given by $\{i \in I \mid g_{I_i}(z^0) \geq -\epsilon_{ls}\}$ with $\epsilon_{ls} > 0$. In the implementation we used $\epsilon_{ls} = 2.0$. The algorithm sSQPa has been designed to set $\lambda^0 \leftarrow \hat{\lambda}^0$. Our numerical experiments have however shown that the solution of (7) is a better choice for λ^0.

Other parameters have been set as follows: $\sigma = 0.95$, $\tau = 0.95$, and $\hat{\tau} = 0.85$. However we have used a different value for σ in the decrease condition $\eta(z^k + \Delta z, \lambda^+) \leq \left(\eta(z^k, \lambda^k)\right)^{1+\sigma/2}$ that appears in the sSQPa algorithm. We have tried several possibilities and conclude that a robust choice for σ in this condition is 0.05.

Test Problem 1: The first test problem is HS113 [5], where degeneracy is due to lack of strict complementarity. This problem has 10 variables and 8 inequality constraints. The performance of sSQPa for three different starting points is given in figure 1. The numerical results show convergence from remote points and fast rate of local convergence. In figure 7 one can see the decrease in $\eta(z, \lambda)$. Table 1 shows that sSQPa is quite competitive with other solvers on this problem in number of iterations (starting from the standard point for this problem).

Test Problem 2: The second test problem is from [2] and it is a modified version of HS46 [5]. Degeneracy in this problem is due to lack of strict complementarity.

Table 1. Performance of different solvers on example 1

Solver	NPSOL	SNOPT	NITRO	LOQO	sSQPa
Iterations	14	32	15	16	9
Objective	24.306	24.306	24.306	24.306	24.306

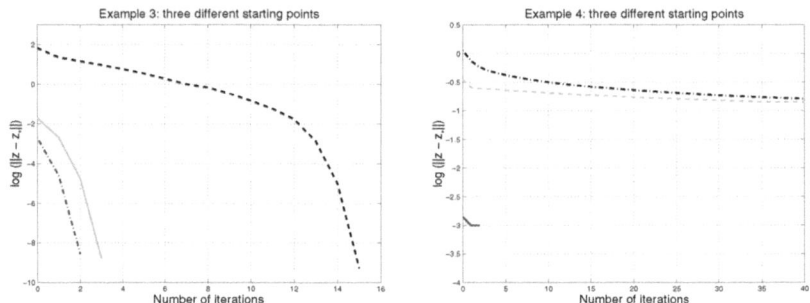

Fig. 2. Convergence from three different starting points for problems 3 and 4

The problem has 5 variables and 3 inequality constraints. Figures 1 and 7 show convergence from remote points and fast rate of local convergence.

In this problem the effect of updating the multiplier using the interior multiplier $\hat{\lambda}$ can be nicely observed. In fact, we can see from table 2 that at the first, tenth, and twelfth iterations the decrease in $\eta(z, \lambda)$ has been poor. In this case sSQPa selects the **else** condition and updates λ^k by $\lambda^k \leftarrow \hat{\lambda}^k$, speeding up the rate of local convergence.

Test Problem 3: The third test problem is also from [2] and it is rank deficient. This problem is a modified version of HS43 [5] and has 4 variables and 4 inequality constraints. We see that sSQPa has converged from the three starting points and has exhibited a fast rate of local convergence (see figures 2 and 7).

Test Problem 4: The fourth test problem is HS13 [5]. It is rank deficient and, furthermore, it does not satisfy the Karush-Kuhn-Tucker and Mangasarian-

Table 2. Performance of sSQPa on example 2. $\text{cond}(H^k)$ is the condition number of the Hessian of the QP, i_{id} is the number of iterations needed by procedure ID0

k	$\phi(z^k)$	$\eta(z^k, \lambda^k)$	i_{id}	$\|\lambda^k\|$	$\|z^k - z^*\|$	$\text{cond}(H^k)$
0	3.33763e+00	7.87595e+00	0	2.14476e-01	1.70244e+00	—
1	3.33763e+00	7.87595e+00	2	5.01797e+00	1.70244e+00	6.82398e+15
2	4.68754e-01	2.43281e+00	0	8.36948e-02	1.34192e+00	$+\infty$
$\vdots$	$\vdots$	$\vdots$	$\vdots$	$\vdots$	$\vdots$	$\vdots$
9	2.74649e-07	1.03603e-03	0	9.99099e-04	3.81399e-02	3.11708e+03
10	2.74649e-07	1.03603e-03	1	0	3.81399e-02	1.20856e+05
11	2.81219e-10	2.02709e-05	0	1.04462e-05	2.18212e-04	$+\infty$
12	2.81219e-10	2.02709e-05	1	0	2.18212e-04	4.09199e+10
13	2.05082e-23	5.54501e-12	0	6.09971e-18	2.93470e-08	$+\infty$

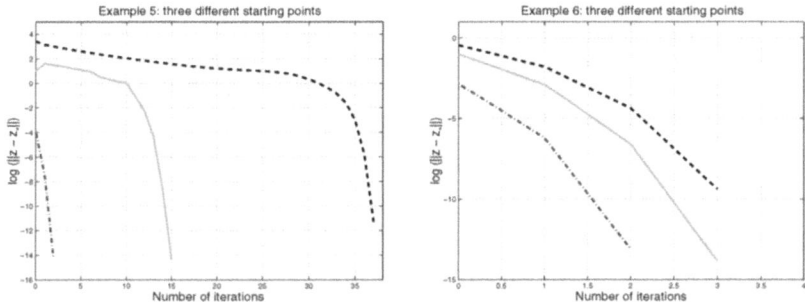

Fig. 3. Convergence from three different starting points for examples 5 and 6

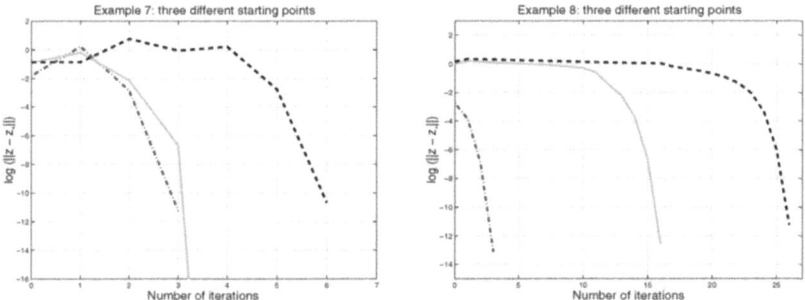

Fig. 4. Convergence from three different starting points for examples 7 and 8

Fromovitz constraint qualifications. It has 2 variables, 1 inequality constraint, and 2 bound constraints.

In figure 2 one can see the convergence behavior from the three starting points and observe that sSQPa approaches a point different from z^* in all of them. Convergence from the first starting point is achieved in two iterations, while convergence for the other two starting points is very slow and the lowest value of $\eta(z, \lambda)$ is 10^{-3} in 40 iterations. Other solvers, among them NPSOL, SNOPT, and NITRO, exhibit a similar behavior for this problem by not converging to the solution.

Test Problem 5: The fifth problem is a modified version of HS100 [5] and it is rank deficient. The problem has 7 variables and 5 inequality constraints and

Table 3. Performance of different solvers on example 8

Solver	NPSOL	SNOPT	NITRO	LOQO	sSQPa
Iterations	2	4	17	23	16
Objective	1.0	1.0	1.000002	1.0	1.0

has the following form:

$$\min \phi(z) = (z_1 - 10)^2 + 5(z_2 - 12)^2 + z_3^4 + 3(z_4 - 11)^2$$
$$+10z_5^6 + 7z_6^2 + z_7^4 - 4z_6z_7 - 10z_6 - 8z_7$$

$$\text{s.t. } g_1(z) = 2z_1^2 + 3z_2^4 + z_3 + 4z_4^2 + 5z_5 - 127 \leq 0,$$
$$g_2(z) = 7z_1 + 3z_2 + 10z_3^2 + z_4 - z_5 - 282 \leq 0,$$
$$g_3(z) = 23z_1 + z_2^2 + 6z_6^2 - 8z_7 - 196 \leq 0,$$
$$g_4(z) = 4z_1^2 + z_2^2 - 3z_1z_2 + 2z_3^2 + 5z_6 - 11z_7 \leq 0,$$
$$g_5(z) = z_1^2 + 1.5z_2^4 + 0.5z_3 + 2z_4^2 + 2.5z_5 - 63.5 \leq 0.$$

Figures 3 and 7 show that the global and local performance of sSQPa is good for this problem.

Test Problem 6: This test problem is considered in [11] and it is rank deficient. It has 2 variables and 2 inequality constraints. There is nothing special to report; the global and local behavior of sSQPa for this problem are fine (see figures 3 and 7).

Test Problem 7: The seventh problem is a quadratic problem with quadratic constraints introduced in [7]. The problem has 3 variables and 6 inequality constraints. The degeneracy is due to lack of strict complementarity. The performance of sSQPa for this problem is shown in figures 4 and 7 and was good both globally and locally.

Test Problem 8: This test problem is HS32 [5] in which strict complementarity does not hold. It has 3 variables, 1 inequality constraint, 1 equality constraint, and 3 bound constraints. Global convergence and fast local convergence for this example can be confirmed from figures 4 and 7. In addition, table 3 gives a comparison between sSQPa and other solvers on this problem for the standard starting point associated with this problem.

Test Problem 9: In this example we have modified problem HS40 by adding (one and two) cuts to the set of constraints. The cuts have made the problem rank deficient. In the case of two cuts, the problem has 4 variables, 2 inequality

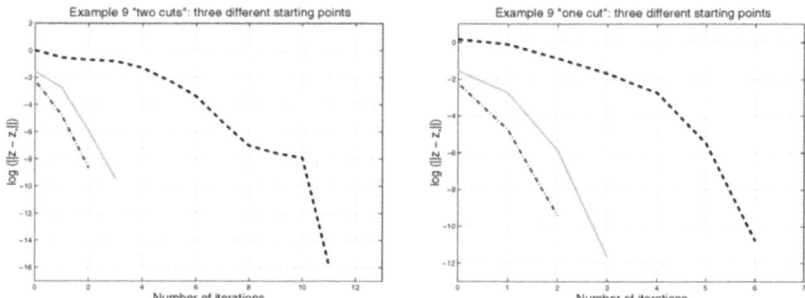

Fig. 5. Convergence from three different starting points for example 9

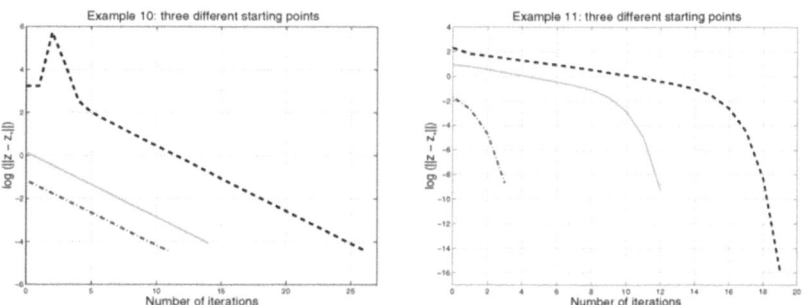

Fig. 6. Convergence from three different starting points for examples 10 and 11

Table 4. Performance of different solvers on example 10

Solver	DONLP2	FilterSQP	LANCELOT	LINF	LOQO	MINOS	SNOPT	sSQPa
$\|z_f - z^*\|$	1.5e-16	5.3e-09	8.7e-07	1.1e-08	1.6e-07	4.8e-06	3.4e-07	1.1e-08
Iters.	4	28	336	28	200	27	3	27

constraints, and 3 equality constraints:

$$\begin{aligned}
\min \phi(z) &= -z_1 z_2 z_3 z_4 \\
\text{s.t. } g_1(z) &= -z_1 z_2 z_3 z_4 + 0.25 &\leq 0, \\
g_2(z) &= -0.5 z_1 z_2 z_3 z_4 + 0.124999 \leq 0, \\
g_3(z) &= z_1^3 + z_2^2 - 1 &= 0, \\
g_4(z) &= z_1^2 z_4 - z_3 &= 0, \\
g_5(z) &= -z_2 + z_4^2 &= 0.
\end{aligned}$$

The one-cut case is generated by omitting the second constraint. The numerical behavior for this example is shown in figures 5 and 8.

Test Problem 10: This problem is taken from [1] and it is rank deficient. It has 3 variables and 4 inequality constraints. We observe that sSQPa exhibits a

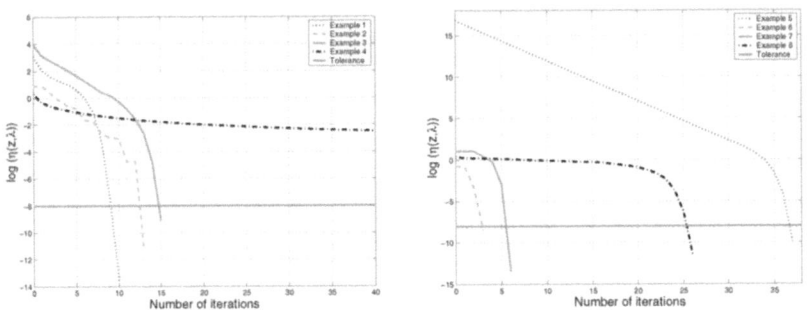

Fig. 7. Performance of sSQPa on examples 1 to 8

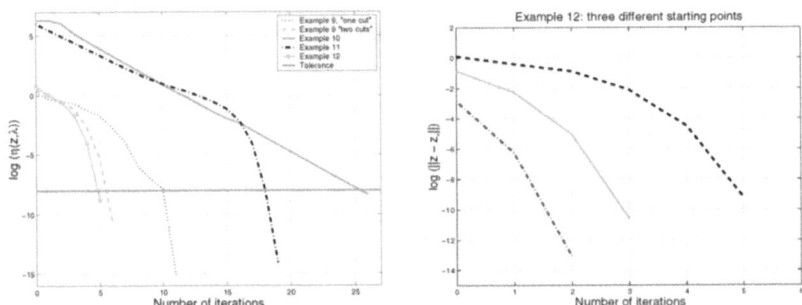

Fig. 8. Performance of sSQPa on examples 9 to 12; convergence from three different starting points for example 12

linear rate of local convergence for this example (see figure 8). In table 4 we have restated the comparison made in [1], listing the distance $\|z_f - z^*\|$ of the final point z_f to the optimal solution z^* and the number of iterations. It is shown in [1] that methods that are based on augmented Lagrangian functions do not perform well on this problem. To some extent the stabilized SQP method has the flavor of augmented Lagrangian methods since the quadratic model of the Lagrangian is augmented by another term involving the stabilization parameter, see the QP problem (3), and this might explain the not so good performance of the sSQPa algorithm on this problem.

Test Problem 11: This problem is a modified version of problem HS43 and example 3, where an equality is included to the set of constraints. It is rank deficient and has the following form:

$$
\begin{aligned}
\min \phi(z) &= z_1^2 + z_2^2 + 2z_3^2 + z_4^2 - 5(z_1 + z_2) - 21z_3 + 7z_4 \\
\text{s.t. } g_1(z) &= z_1^2 + z_2^2 + z_3^2 + z_4^2 + z_1 - z_2 + z_3 - z_4 - 8 & \leq 0, \\
g_2(z) &= z_1^2 + 2z_2^2 + z_3^2 + 2z_4^2 - z_1 - z_4 - 10 & \leq 0, \\
g_3(z) &= 2z_1^2 + z_2^2 + z_3^2 + 2z_1 - z_2 - z_4 - 5 & \leq 0, \\
g_4(z) &= -z_2^3 - 2z_1^2 - z_4^2 - z_1 + 3z_2 + z_3 - 4z_4 - 7 & = 0.
\end{aligned}
$$

The numerical behavior of sSQPa for this problem can be seen from figures 6 and 8 and it is characterized by fast local convergence for the three different starting points.

Test Problem 12: This test problem is a modified version of problem 6. It has an unique minimizer at the origin. It is also rank deficient and has the following form:

$$\min \phi(z) = z_1$$
$$\text{s.t.} \ \ g_1(z) = (z_1 - 2)^2 + z_2^2 - 4 \ \ \leq 0,$$
$$g_2(z) = -(z_1 - 4)^2 - z_2^2 + 16 \leq 0.$$

The optimal set of the Lagrange multipliers is

$$\mathcal{S}_\lambda = \{(\alpha, \alpha/2) \mid \alpha \geq 0\},$$

which is clearly unbounded, implying that the MFCQ does not hold at the solution. The algorithm sSQPa exhibits a fast local rate of convergence on this example. Figure 8 shows the performance of sSQPa on this example.

4.2   Performance of Procedure ID0

The aim of this subsection is to give some insight on the numerical behavior of procedure ID0 within algorithm sSQPa on the test problems introduced in subsection 4.1. The goal is to see how close to z^* the procedure ID0 is able to correctly identify the active constraints and its partition into strong and weak active constraints. We report in table 5 the value of $\|z - z^*\|$ from which these identifications is always correct.

With the exception of example 2, we used the results of the farthest away starting point. In fact, procedure ID0 does not detect the active set correctly in example 2 for the second and third starting points. In addition, procedure ID0 does not detect the active set correctly for any of the three starting points of example 7. In general, we can say that procedure ID0 does a good job identifying the active constraints and its partition into strong and weak.

---

[4] For the first and second starting points it is obtained a correct partitioning of $\mathcal{A}(z, \lambda)$ into $\mathcal{A}_+$ and $\mathcal{A}_0$, while for the third starting point g_4 is incorrectly in $\mathcal{A}_0$ from the tenth iteration until the end.

[5] The second cut is perturbed a little so that its value at the solution is 10^{-7} (see problem 9). Procedure ID0 identifies the two constraints in $\mathcal{A}(z, \lambda)$ from the initial starting point until $\|z - z^*\| = 1.2e - 08$, and then excludes g_2 incorrectly from $\mathcal{A}(z, \lambda)$.

[6] At the first iteration $\mathcal{A}_+ = \mathcal{A}(z, \lambda) = \{1, 2\}$, then in the following 4 iterations g_2 is excluded from $\mathcal{A}(z, \lambda)$ until the last iteration (the fifth iteration) at which g_2 returns to $\mathcal{A}(z, \lambda)$ with a value of $-3.04e - 09$.

Table 5. Detection of the correct active constraints and its partition into strong and weak

Test Problem	$\mathcal{A}(z,\lambda)$	$\|z-z^*\|$	$\mathcal{A}_+$	$\mathcal{A}_0$	$\|z-z^*\|$
1	$\{1,2,3,4,5,7\}$	5.3e-01	$\mathcal{A}_+ = \mathcal{A}(z,\lambda)$	$\mathcal{A}_0 = \emptyset$	6.2e-05
2	$\{1,2,3\}$	1.4e-05	$\mathcal{A}_+ = \emptyset$	$\mathcal{A}_0 = \mathcal{A}(z,\lambda)$	2.4e-10
3	$\{1,3,4\}$	1.4e-01	$\mathcal{A}_+ = \{1,3\} \neq \mathcal{B}_+$	$\mathcal{A}_0 = \{4\} \neq \mathcal{B}_0$ [1]	1.4e-01
4	$\{1,3\}$	8.7e-01	$\mathcal{A}_+ = \mathcal{A}(z,\lambda)$	$\mathcal{A}_0 = \emptyset$	6.3e-01
5	$\{1,4,5\}$	1.1e+00	$\mathcal{A}_+ = \mathcal{A}(z,\lambda)$	$\mathcal{A}_0 = \emptyset$	9.7e-04
6	$\{1,2\}$	3.6e-01	$\mathcal{A}_+ = \mathcal{A}(z,\lambda)$	$\mathcal{A}_0 = \emptyset$	4.3e-05
7					
8	$\{2,3\}$	6.3e-01	$\mathcal{A}_+ = \mathcal{A}(z,\lambda)$	$\mathcal{A}_0 = \emptyset$	6.3e-01
9 "two cuts" [2]	$\{1\}$	1.2e-08	$\mathcal{A}_+ = \mathcal{A}(z,\lambda)$	$\mathcal{A}_0 = \emptyset$	1.2e-08
9 "one cut"	$\{1\}$	7.6e-01	$\mathcal{A}_+ = \mathcal{A}(z,\lambda)$	$\mathcal{A}_0 = \emptyset$	7.6e-01
10	$\{1,2,3,4\}$	4.3e-02	$\mathcal{A}_+ = \mathcal{A}(z,\lambda)$	$\mathcal{A}_0 = \emptyset$	2.1e-02
11	$\{1,3\}$	2.1e-01	$\mathcal{A}_+ = \mathcal{A}(z,\lambda)$	$\mathcal{A}_0 = \emptyset$	2.1e-01
12	$\{1,2\}$	6.4e-10	$\mathcal{A}_+ = \mathcal{A}(z,\lambda)$	$\mathcal{A}_0 = \emptyset$	6.4e-10 [3]

4.3   Problems without objective Function

We have tested the ability of the sSQPa method to find feasible points for feasible degenerate problems and to find least infeasible points for infeasible degenerate problems. For this purpose the objective function and its derivatives were set to zero in the algorithm. No globalization scheme was used. We have tried four possibilities for the function $\eta(z,\lambda)$ used in the stopping criterion:

$$\eta_1(z,\lambda_I) = \left\|\begin{bmatrix} \min(\lambda_I, -g_I(z)) \\ g_E(z) \end{bmatrix}\right\|, \quad \eta_2(z,\lambda) = \left\|\begin{bmatrix} \nabla_z \mathcal{L}(z,\lambda) \\ \min(\lambda_I, -g_I(z)) \\ g_E(z) \end{bmatrix}\right\|,$$

$$\eta_3(z,\lambda) = \left\|\begin{bmatrix} g_{\mathcal{A}(z,\lambda)}(z) \\ g_E(z) \end{bmatrix}\right\|, \quad \eta_4(z) = \left\|\begin{bmatrix} \max(g_I(z),0) \\ g_E(z) \end{bmatrix}\right\|.$$

Note that $\eta_2(z,\lambda)$ is the one that is used in computing stationary points and that $\eta_4(z)$ represents a measure of the true feasibility.

We ran the 12 feasible problems introduced in section 4.1. In addition, we have designed 2 more infeasible problems that will be described later on in this section. For each of these problems we considered three different starting points. The stopping criterion was also $\eta(z,\lambda) \leq 10^{-8}$. The initial multipliers have been obtained by

$$\lambda^0 = \begin{bmatrix} \max(0, g_I(z^0)) \\ g_E(z^0) \end{bmatrix}.$$

The overall results are given in table 6, where we report the average number of iterations and the number of wins out of 57 trials for each of the four types of $\eta(z,\lambda)$ given above. We observe that $\eta_1(z,\lambda)$ and $\eta_4(z)$ seem to be most efficient choices. With the exception of the second and third starting points of example 8,

Table 6. Performance of sSQPa for finding feasible or infeasible points

	$\eta_1(z, \lambda_I)$	$\eta_2(z, \lambda)$	$\eta_3(z, \lambda)$	$\eta_4(z)$
Iteration average	10.18	10.63	12.95	9.23
Number of wins	50	48	25	52

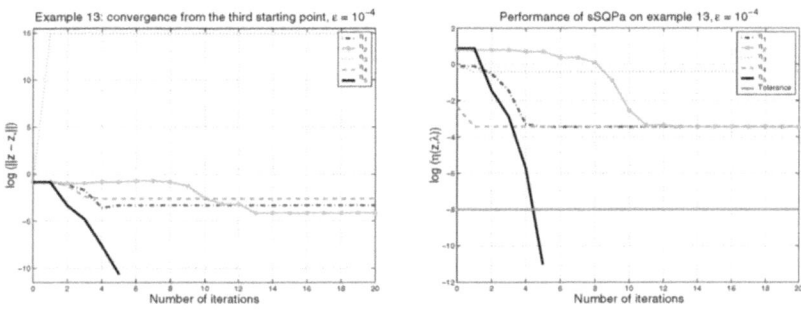

Fig. 9. Convergence for example 13: $\epsilon = 10^{-4}$

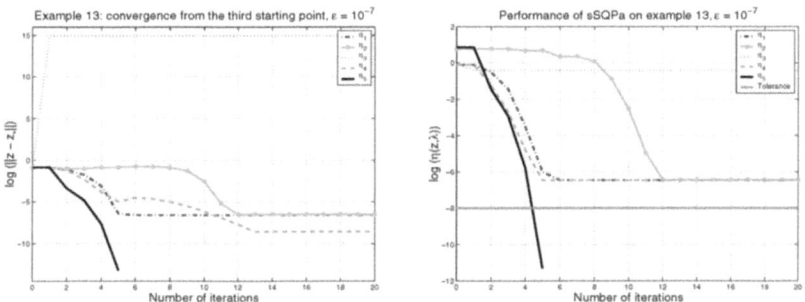

Fig. 10. Convergence for example 13: $\epsilon = 10^{-7}$

and of the three starting points of example 11, sSQPa with η_4 has successfully converged to feasible points.

In general, the feasible point computed changes with the starting point z^0 and the choice of $\eta(z, \lambda)$. The sSQPa method also worked well for the two cases of problem 9 and problem 12, with the particularity that the feasible point computed was found to be also stationary.

Next, we introduce two infeasible test problems, where at the least infeasible points the gradients of the nearby active constraints are linearly dependent.

Test Problem 13: This test problem is a modification of both of problems 6 and 12, and has the following form:

$$g_1(z) = (z_1 - (2 + \epsilon))^2 + z_2^2 - 4 \quad \leq 0,$$
$$g_2(z) = -(z_1 - 4)^2 - z_2^2 + (4 + \epsilon)^2 \leq 0.$$

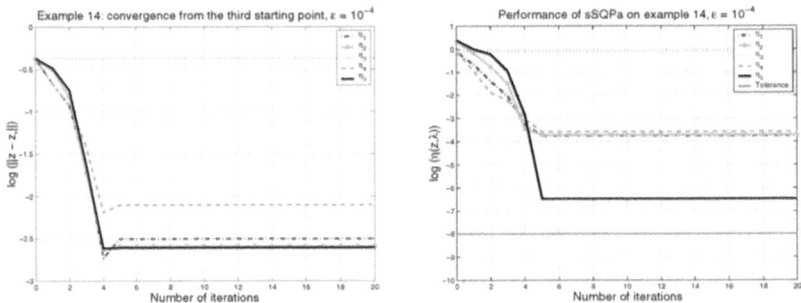

Fig. 11. Convergence for example 14: $\epsilon = 10^{-4}$

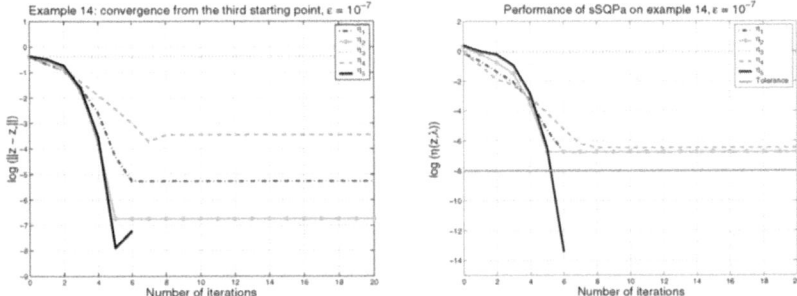

Fig. 12. Convergence for example 14: $\epsilon = 10^{-7}$

If ϵ is set to zero, the problem becomes feasible and rank deficient at the solution. The problem is infeasible for small positive values of ϵ.

In the computations, we have tested this example for $\epsilon = 10^{-4}$ and $\epsilon = 10^{-7}$. Again, the sSQPa method has been applied to these two instances with the objective function and its derivatives set to zero and without any globalization scheme.

The results are given in figures 9 and 10 for the two instances ($\epsilon = 10^{-4}$ and $\epsilon = 10^{-7}$) and for the four choices of $\eta(z, \lambda)$, in terms of the distance to the least-squares minimizer z^* of the constraints obtained by **Matlab** and also in terms of $\eta(z, \lambda)$. Here we tested another function $\eta(z, \lambda)$ for the stopping criterion:

$$\eta_5(z, \lambda) = \left\| \sum_{i \in \mathcal{A}(z,\lambda) \cup E} g_i(z) \nabla g_i(z) \right\|.$$

Test Problem 14: Test problem 14 is also infeasible, defined by an hyperplane and a circle:

$$\begin{aligned} g_1(z) &= z_1 + \epsilon & \leq 0, \\ g_2(z) &= (z_1 - (1 + \epsilon))^2 + z_2^2 - 1 \leq 0. \end{aligned}$$

The results are given in figures 11 and 12. The main conclusion that we can draw from these two test problems is that the stabilized SQP with constraint identification (algorithm sSQPa) was quite effective to determine least infeasible points with nearby rank deficiency. Among the five measures of least infeasibility, $\eta_5(z, \lambda)$ seems to be the most efficient.

References

[1] M. ANITESCU, *Degenerate nonlinear programming with a quadratic growth condition*, SIAM J. Optim., 10 (2000) 1116–1135.

[2] F. FACCHINEI, A. FISCHER, AND C. KANZOW, *On the accurate identification of active constraints*, SIAM J. Optim., 9 (1998) 14–32.

[3] W. W. HAGER, *Stabilized sequential quadratic programming*, Comput. Optim. and Appl., 12 (1999) 253–273.

[4] M. HEINKENSCHLOSS, M. ULBRICH, AND S. ULBRICH, *Superlinear and quadratic convergence of affine-scaling interior-point Newton methods for problems with simple bounds without strict complementarity assumption*, Math. Programming, 86 (1999) 615–635.

[5] W. HOCK AND K. SCHITTKOWSKI, *Test examples for nonlinear programming codes*, Lecture Notes in Economics and Math. Systems, 187, Springer Verlag, Berlin, 1981.

[6] D.-H. LI AND L. QI, *A stabilized SQP method via linear equations*, Technical report, Mathematics Department, University of New South Wales, 2000.

[7] A. V. DE MIGUEL AND W. MURRAY, *Generating optimization problems with global variables*, Technical Report TR/SOL 01–3, Management Science & Eng., Stanford Univ., 2001.

[8] D. RALPH AND S. J. WRIGHT, *Superlinear convergence of an interior-point method despite dependent constraints*, Math. Oper. Res., 25 (2000) 179–194.

[9] L. N. VICENTE AND S. J. WRIGHT, *Local convergence of a primal–dual method for degenerate nonlinear programming*, Preprint 00–05, Department of Mathematics, University of Coimbra, 2000 (Revised May 2001).

[10] S. J. WRIGHT, *Modifying SQP for degenerate problems*, Preprint ANL/MCS–P699-1097, Argonne National Laboratory, 1998, (Revised June, 2000).

[11] S. J. WRIGHT, *Superlinear convergence of a stabilized SQP method to a degenerate solution*, Comput. Optim. and Appl., 11 (1998) 253–275.

[12] S. J. WRIGHT, *Constraint identification and algorithm stabilization for degenerate nonlinear programs*, Preprint ANL/MCS–P865-1200, Argonne National Laboratory, 2000 (Revised Nov. 2001).

A New Method for the Global Solution of Large Systems of Continuous Constraints

Mark S. Boddy[1] and Daniel P. Johnson[2]

[1]Adventium Labs
105 Mill Place, 111 Third Ave. S., Minneapolis, MN 55401, USA
phone: 651-442-4109
mark.boddy@adventiumlabs.org
[2]Honeywell Laboratories
3660 Technology Drive, Minneapolis, MN 55418, USA
phone: 612-951-7403
daniel.p.johnson@honeywell.com

Abstract. Scheduling of refineries is a hard hybrid problem. Application of the Constraint Envelope Scheduling (CES) approach required development of the Gradient Constraint Equation Subdivision (GCES) algorithm, a novel global feasibility solver for the large system of quadratic constraints that arise as subproblems. We describe the implemented solver and its integration into the scheduling system. We include discussion of pragmatic design tradeoffs critically important to achieving reasonable performance.

1   Introduction

We are conducting an ongoing program of research on modeling and solving complex hybrid programming problems (problems involving a mix of discrete and continuous variables), with the end objective of implementing improved finite-capacity schedulers for a wide variety of different application domains, in particular manufacturing scheduling in the refinery and process industries.

In this report we present an algorithm that will either find a feasible solution or establish global infeasibility for a quadratic system of continuous equations generated as a subproblem in the course of solving finite-capacity scheduling problems in a petroleum refinery domain.

2   Motivation

Prediction and control of physical systems involving complex interactions between a continuous dynamical system and a set of discrete decisions is a common need in a wide variety of application domains. Effective design, simulation and control of such hybrid systems requires the ability to represent and manipulate models including both discrete and continuous components, with some interaction between those components.

C. Bliek et al. (Eds.): COCOS 2002, LNCS 2861, pp. 142-156, 2003.

Refinery operations have traditionally been broken down as follows (see figure 1):

- Crude receipts and crude blending, which encompasses everything from the initial receipt of crude oils through to the crude charge provided to the crude distillation unit (CDU).
- The refinery itself, involving processing the crude through a variety of processing units and intermediate tanks to generate blendstocks, which are used as components to produce the materials sold by the refinery.
- Product blending, which takes the blendstocks produced by the refinery and blends them together so as to meet requirements for product shipments (or liftings).

In Figure 1, crude receipts and blending encompass managing the crude tanks (depicted as cylinders) and crude disillation unit (rounded rectangle) to the left of the leftmost dotted line. The processing units and intermediate inventory of the refinery itself are represented similarly, appearing between the dotted green lines. Product blending and lifting encompasses the blenders (rectangles), tanks, and shipment points (arrows) on the right of the figure.

Refinery operators have recognized for years that solving these problems independently can result in global solutions that are badly suboptimal and difficult to modify rapidly and effectively in response to changing circumstances. Standard practice is to aggregate the production plans across multiple periods, but current solvers are able to handle only a few planning periods.

Constructing a model of refinery operations suitable for scheduling across the whole refinery requires the representation of asynchronous events, time-varying continuous variables, and mode-dependent constraints. In addition, there are key quadratic interrelationships between volume, rate, and time, mass, volume and specific gravity, and among tank volumes, blend volumes and blend qualities. This leads to a system containing quadratic constraints with equalities and inequalities.

A refinery planning problem may involve hundreds or thousands of such variables and equations. The corresponding scheduling problem may involve thousands or tens of thousands of variables and constraints. Only recently has the state of the art (and, frankly, the state of the computing hardware) progressed to the point where scheduling the whole refinery on the basis of the individual processing activities themselves has entered the realm of the possible.

Scheduling the whole refinery at once involves the following discrete decisions:

- CDU mode (determines allowable non-zero input and output streams)
- Other processing-unit modes
- Product blends (ordering, material, destination tank).
- Product liftings (source tank).

as well as the following continuous decisions:

- Crude feed proportions for each crude charge
- Process unit settings (CDU, desulphurizer, various other)
- Stream percentages (how much goes to which alternate destination, including intermediate tanks).

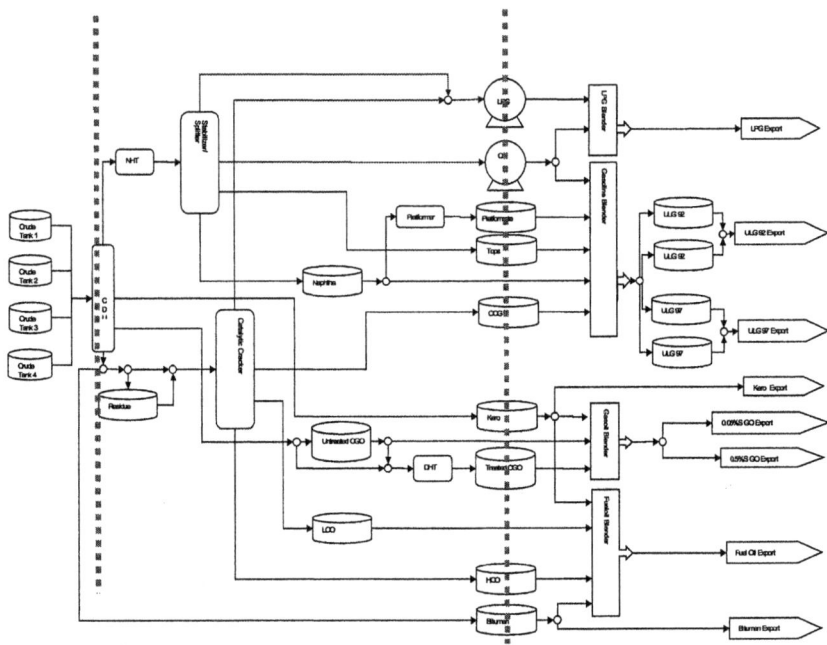

Fig. 1. Example of refinery operation, divided into traditional planning and scheduling operational areas

Intermediate tanks are one of the more interesting issues, because of the requirements for quality tracking (and quality effects on unit operations), and the consequent opportunities to use intermediate tanks for blending properties over time by holding material in the tanks over some period of time.

The GCES solver described in the rest of this paper has been implemented as one component of a scheduling system that addresses the whole-refinery scheduling problem described in this section. The GCES solver is a critical component of that scheduler, without which the problem could not be expressed or solved in an acceptable period of time.

2.1   Constraint Envelope Scheduling

The scheduling process for a refinery involves a complex interaction between discrete choices (resource assignments, sequencing decisions), a continuous temporal model, and continuous processing and property models. Discrete choices enforce new constraints on the continuous models. Those constraints may be inconsistent with the current model, thus forcing backtracking in the discrete domain, or if consistent, may serve to constrain choices for discrete variables not yet assigned.

Constraint envelope scheduling [1] is a least-commitment approach to constraint-based scheduling, in which restrictions to the schedule are made only as necessary. It uses an explicit representation of scheduling decisions (e.g., the decision to order two activities to remove a resource conflict) which permits identification of the source, motivation, and priority of scheduling decisions, when conflicts arise in the process of schedule construction or modification.

The resulting approach supports a continuous range of scheduling approaches, from completely interactive to completely automatic. We have built systems across this entire range of interaction styles, for a variety of applications including aircraft avionics [2], image data analysis and retrieval [3], spacecraft operations scheduling, and discrete and batch manufacturing [6], among others, with problem sizes ranging up to 30,000 activities and 140,000 constraints.

2.2   Hybrid Solvers for Mixed Variable Problems

In [4], we presented an architecture for the solution of hybrid constraint satisfaction or constrained optimization problems, using a communicating set of discrete and continuous solvers. Integration into this architecture imposes a set of requirements on both the discrete and continuous solvers. The most difficult requirement, once quadratic or nonlinear continuous constraints are present, is that the continuous solver must be capable of efficiently establishing the global feasibility or infeasibility of the continuous problem.

In past research, we have built hybrid constraint solvers with linear solvers customized for temporal models, and with general linear programming solvers, specifically the CPLEX linear programming libraries.

More recently, we have implemented the GCES quadratic solver that uses CPLEX as a subroutine, and integrated that as well. The result is that we have now implemented a hybrid solver, using the architecture described in [4], that combines a discrete solver with a continuous solver capable of handling quadratic (and thus through rewriting, arbitrary polynomial and rational) constraints.

The continuous GCES solver described in this paper is specifically designed to function effectively as a component of the overall hybrid solver. Specifically, the GCES solver has the following characteristics:

Detection of Global Infeasibility. As the hybrid solver progresses by making discrete decisions, the addition of continuous constraints may result in an infeasible continuous problem, necessitating backtracking in the discrete problem. Rapid and effective determination of infeasibility is thus a critical capability. This efficiency is achieved in our solver through the use of specialized propagation methods and approximate solutions (e.g., generating a solution for only the linear part of the problem).

Incremental Addition of Constraints. The solution process proceeds through the incremental solution of a hybrid Constraint Satisfaction Problem (CSP), at each step attempting to establish whether the current partial assignment is feasible (and thus potentially extendible to a complete solution). This process involves the repeated addition (and, on backtracking, deletion) of small numbers of constraints to a large store of constraints that is otherwise unchanged. Given the scale of the problem we addressed (see Section 3.5) and the number of times the continuous solver is called in the solution process, incremental modification of the current partial solution was crucial to achieving acceptable performance.

2.3   Related Work

In the past few years, hybrid programming has become a very active research area. Here, we touch only upon a few of the basic approaches.

The most common approach has been the use of Mixed Integer Linear Programming (MILP). Typically, schedules in an MILP approach are modeled using multi-period aggregation. An increasing body of work (e.g. [13], [7]) shows that simple integer models lack efficiency and performance (though not, of course, expressive power).

Newer approaches seek to integrate constraint or logic programming with continuous solvers, typically linear programming. A variety of approaches are being explored, differing on such design issues as how the discrete and continuous models are connected or integrated (variables, constraints, functional decomposition), the level of decomposition (how often solving effort moves between the discrete and continuous models), and the solver's control structure (is it an extension of Logic Programming, a simple branching structure mediating the solution of different continuous or mixed-integer models, or something else entirely?).

Heipcke introduces variables that associate the two domains [7]. Hooker [8] has been developing solution methods for problems expressed as a combination of logical formulas and linear constraints. Lee and Grossman [11] have introduced disjunctive programming. Van Hentenryck [14] uses box consistency to solve continuous problems.

One of the keys to providing an integrated solver for hybrid models is the proper handling of the tradeoffs and interactions between the discrete and continuous domains. Heipcke's thesis provides a very flexible system for propagating across both. The approach outlined by Boddy et. al. [4] uses discrete search to direct the continuous solver.

3   Subdivision Search

The Gradient Constraint Equation Subdivision (GCES) solver accepts systems of quadratic equations, quadratic inequalities, and variable bounds, and either will find a feasible solution or will establish the global infeasibility of the system, within the normal limits of numerical conditioning.

3.1   Overview

The method presented here is based on adaptively subdividing the ranges of the variables until the equations are sufficiently linear on the subdivided region that linear methods are sufficient to either find a feasible point, or to prove the absence of any solution to the original equation.

Within each subdivided region, the method uses two local algorithms, one to determine if there is no root within the region, and one to determine if a feasible point can be found. If both methods fail, then that region is subdivided again. The algorithm terminates once a feasible point is found, or all regions have been proven infeasible.

Kantorovich's theorem on Newton's method establishes that for small enough regions, one can use linear methods to either find a feasible solution, or to prove infeasibility. R. E. Moore [12] proposed this as the basis for a global solver in the context of the field of interval arithmetic. Cucker and Smale [5] have proven rigorous performance bounds for such an algorithm with a grid search.

The GCES infeasibility test uses a enveloping linear program known as the Linear Program with Minimal Infeasibility (LPMI), which uses one-sided bounds for the upper and lower limits of the gradients of the equations within the region. Its infeasibility rigorously establishes the infeasibility of the original nonlinear constraints. GCES currently uses a stabilized version of Successive Linear Programming (SLP) to determine feasibility. Alternatives are under current investigation.

We also have introduced the use of continuous constraint propagation methods to refine the variable bounds as the ranges are subdivided. This technique interacts well with the local linearization methods as the reduced bounds often improve the efficiency of the linearizations. The Newton system by Van Hentenryck [12] is a successful example of a global solver that involves the use of only subdivision and propagation, without the use of an enveloping approximation to establish infeasibility or the use of linearization to speed convergence.

3.2   Basic Algorithm

Let $f(x): \mathfrak{R}^n \to \mathfrak{R}^m$ be a quadratic function of the form

$$f_k(x) = C_k + \sum_i A_{ki} x_i + \sum_{ij} B_{kji} x_j x_i \ . \tag{1}$$

With an abuse of notation, we shall at times write the functions as $f(x) = C + Ax + Bxx$.

We then put upper and lower bounds $lb \in \mathfrak{R}^m, ub \in \mathfrak{R}^m$ on the functions, and lower and upper bounds $u^0 \in \mathfrak{R}^n, v^0 \in \mathfrak{R}^n$ on the variables. (These bounds are allowed to be equal to express equalities.) The problem we wish to solve will have the form

$$P0 = \left\{ x: \ u^0 \le x \le v^0, \ \ lb \le f(x) \le ub \right\} \tag{2}$$

We define the *infeasibility* of a constraint k at a point x to be

$$\Delta_k(x) = \max\left((f_k(x) - ub_k)_+, (lb_k - f_k(x))_+ \right) \tag{3}$$

where the positive and negative parts are defined as $(x)_+ = \max(x,0)$ and $(x)_- = \min(x,0)$.

The *infeasibility* (resp. *max infeasibility*) of the system at a point x is

$$\Delta(x) = \sum_k \Delta_k(x) , \text{(resp. } \Delta^\infty(x) = \max_k (\Delta_k(x)) \text{)}. \tag{4}$$

In the course of solving the problem above, we will be solving a sequence of subsidiary problems. These problems will be parameterized by a trial solution $\bar{x}$ and a set of point bounds $\{u, v\} : u \le \bar{x} \le v$.

Given the point bounds, we define the gradient bounds

$$F = (B)_+ u + (B)_- v, \quad G = (B)_+ v + (B)_- u \tag{5}$$

(where the positive and negative parts are taken element-wise over the quadratic tensor) so that whenever $u \le x \le v$ we will have $F \le Bx \le G$.

The *gradient range* is given by

$$\delta G(u, v) = G - F = |B|(v - u) \tag{6}$$

and the *maximum infeasibility* is given by

$$\Delta(u, v) = \sum_{ki} (G_{ki} - F_{ki})(v_i - u_i) = |B|(v - u)(v - u) \tag{7}$$

The centered representation of a function relative to a given trial solution $\bar{x}$ is

$$f(x) = \bar{C} + \bar{A}(x - \bar{x}) + B(x - \bar{x})(x - \bar{x}) \tag{8}$$

where $\bar{A} = A + (B + B^*)\bar{x}$, $\bar{C} = C + A\bar{x} + B\bar{x}\bar{x}$, (B^* is the transponse of B).

By also defining $\bar{u} = u - \bar{x}$, $\bar{v} = v - \bar{x}$, $\bar{F} = F - B\bar{x}$, and $\bar{G} = G - B\bar{x}$ the bounding inequalities will be equivalent to the centered inequalities

$$\bar{u} \le x - \bar{x} \le \bar{v}$$
$$\bar{F} \le B(x - \bar{x}) \le \bar{G} \tag{9}$$

In order to develop our enveloping linear problem, we then bound the quadratic equations on both sides by decomposing $x - \bar{x}$ into two nonnegative variables $z, w \ge 0$, $z - w = x - \bar{x}$.

As a result, if $x - \bar{x} \le 0$, we have $-w \le (x - \bar{x}) = (x - \bar{x})_- \le 0 = (x - \bar{x})_+ \le z$, and if $x - \bar{x} \ge 0$, we have $-w \le (x - \bar{x})_- = 0 \le (x - \bar{x})_+ = (x - \bar{x}) \le z$. In both cases we have $-w \le (x - \bar{x})_- \le 0 \le (x - \bar{x})_+ \le z$, which we apply to the quadratic problem to get

$$\bar{F}z - \bar{G}w \le \bar{F}(x - \bar{x})_+ + \bar{G}(x - \bar{x})_-$$
$$\le B(x - \bar{x})(x - \bar{x}) \le \tag{10}$$
$$\bar{G}(x - \bar{x})_+ + \bar{F}(x - \bar{x})_- \le \bar{G}z - \bar{F}w .$$

The subsidiary problems are then

A)   the basic quadratic feasibility problem $P0$ of equation (2).
B)   the basic quadratic problem with the bounding inequalities

$$Pbd(u, v) = \{x : \ u \le x \le v, \ lb \le f(x) \le ub\} \tag{11}$$

C) the linearization problem with the bounding inequalities included

$$LLP(\overline{x}, u, v) = \left\{ x : \quad \overline{u} \le x - \overline{x} \le \overline{v}, \quad lb \le \overline{C} + \overline{A}(x - \overline{x}) \le ub \right\} \tag{12}$$

D) the enveloping linear programming minimal infeasibility (LPMI) problem

$$LPMI(\overline{x}, u, v) = \left\{ x : \exists z, w : \begin{array}{l} \min \Sigma (G - F)(z + w) \\ \overline{u} \le x - \overline{x} \le \overline{v} \\ lb \le \overline{C} + \overline{A}(x - \overline{x}) + \overline{G}z - \overline{F}w \\ ub \ge \overline{C} + \overline{A}(x - \overline{x}) + \overline{F}z - \overline{G}w \\ x - \overline{x} = z - w \\ z + w \le \max(|\overline{v}|, |\overline{u}|) \\ z, w \ge 0 \end{array} \right\} \tag{13}$$

We note here two properties of the LPMI:

- $Pbd(u, v) \subseteq LPMI(\overline{x}, u, v)$, which justifies the use of the LPMI to establish infeasibility of the quadratic system).
- If $x' \in LPMI(\overline{x}, u, v)$ is a solution of the LPMI, then $\Delta(x') \le \Delta(u, v)$, which justifies the use of the term *maximum infeasibility*.

As we search for a solution for a problem $P0$, we split the region up into nodes, each of which is given by a set of bounds $\{u, v\}$.

For any problem, the theory of Newton's method ([5] pp128) tells us that there is a constant $\theta > 0$ such that for any $\Delta(u, v) < \theta$, if $Pbd(u, v)$ is feasible then the sequence of trial solutions generated by the linearization, $\overline{x}^{m+1} = LLP(\overline{x}^m, u, v)$, will converge to a solution $x^* = Pbd(u, v)$.

On the other hand there is a constant $\theta > 0$ such that for any $\Delta(u, v) < \theta$, if $Pbd(u, v)$ is infeasible then $LPMI(\overline{x}, u, v)$ will be infeasible.

So an abstract view of the algorithm can be presented:

1) Among the current node candidates, choose the node $\{u, v\}$ with initial trial solution $u \le \overline{x} \le v$ which has the minimal *max infeasibility* $\Delta^\infty(\overline{x})$.
2) Use bound propagation through the constraints to find refined bounds $\{u', v'\} : u \le u', v' \le v$. If the resulting bounds are infeasible, declare the node infeasible.
3) Iterate $\overline{x}^{m+1} = LLP(\overline{x}^m, u, v)$ until the values either converge to a feasible point or fail to converge sufficiently fast. If the values converge to a feasible point, declare success.
4) Evaluate $x \in LPMI(\overline{x}, u, v)$ for infeasibility, if so, declare the node infeasible.
5) If (2), (3), (4) are all inconclusive for the node, subdivide the node into smaller regions by choosing a variable and subdividing the range of that vari-

able to generate multiple subnodes. Project the trial solution into each region, and try again.

3.3   Propagation

We currently implement two classes of propagation rules, chosen for their simplicity and speed. The first applies to any linear or quadratic function, the second is for so-called "mixing equations".

For the general propagation method, suppose we have the two-sided quadratic function inequality

$$lb_k \leq f_k(x) = C_k + \sum_i A_{ki}x_i + \sum_{ij} B_{kji}x_jx_i \leq ub_k \tag{14}$$

a set of point bounds $u \leq x \leq v$, and we wish to refine the bounds for a variable x_J. We rearrange the inequalities into the form

$$lb_k - C_k - \sum_{i \neq J} A_{ki}x_i - \sum_{i \neq J, j \neq J} B_{kji}x_jx_i \leq \left(A_{kJ} + \sum_{i \neq J}(B_{kJi} + B_{kiJ})x_i + B_{JJ}x_J \right)x_J$$
$$\leq ub_k - C_k - \sum_{i \neq J} A_{ki}x_i - \sum_{i \neq J, j \neq J} B_{kji}x_jx_i \tag{15}$$

We then define quantities

$$\gamma = \left(A_{kJ} + \sum_{i \neq J}(B_{kJi} + B_{kiJ})x_i + B_{JJ}x_J \right)x_J$$
$$\beta = \left(A_{kJ} + \sum_{i \neq J}(B_{kJi} + B_{kiJ})x_i + B_{JJ}x_J \right) \tag{16}$$

By applying the current bounds $u \leq x \leq v$, and applying standard interval arithmentic techniques (see Kearfott [10]) we can find bounds $\{\gamma_0, \gamma_1\}$ and $\{\beta_0, \beta_1\}$ such that

$$\gamma_0 \leq lb_k - C_k - \sum_{i \neq J} A_{ki}x_i - \sum_{i \neq J, j \neq J} B_{kji}x_jx_i$$
$$ub_k - C_k - \sum_{i \neq J} A_{ki}x_i - \sum_{i \neq J, j \neq J} B_{kji}x_jx_i \leq \gamma_1 \tag{17}$$

and

$$\beta_0 \leq \left(A_{kJ} + \sum_{i \neq J}(B_{kJi} + B_{kiJ})x_i + B_{JJ}x_J \right) \leq \beta_1 \tag{18}$$

From the equations and definitions (15)-(18), we see

$$x_J = \frac{\gamma}{\beta} \text{ where } \beta_0 \le \beta \le \beta_1 \text{ and } \gamma_0 \le \gamma \le \gamma_1 \qquad (19)$$

We find bounds $u'_J \le x_J \le v'_J$ again through standard interval arithmetic methods.

A common equation for refinery and chemical processes is a mixing equation of the form

$$x_0(y_1 + y_2 + \ldots + y_n) = x_1 y_1 + x_2 y_2 + \ldots + x_n y_n \qquad (20)$$

where $y_1 \ge 0, y_2 \ge 0, \ldots, y_n \ge 0$.

Here we are given materials indexed by $\{1,2,\ldots,n\}$, and recipe sizes (or rates, or masses, or ratios, whatever mixing unit of interest) $\{y_1, y_2, \ldots, y_n\}$ for a mix of the materials. The equation represents a simple mixing model for a property x_0 of the resulting mix given values $\{x_1, x_2, \ldots, x_n\}$ for the property of each of the materials.

Since the resulting property will be a convex linear combination of the individual properties, if we are given bounds $a_1 \le x_1 \le b_1, a_2 \le x_2 \le b_2, \ldots, a_n \le x_n \le b_n$ then $\min(a_1, a_2, \ldots, a_n) \le x_0 \le \max(b_1, b_2, \ldots, b_n)$.

3.4   Subdivision Strategies

A critical factor in the performance of the solver is the choice of variable to subdivide. We evaluated nine different strategies. The intent of all the strategies considered was to reduce the amount of nonlinearity within the subdivided regions. The better strategies were considered to be those that required fewer subnodes to be generated and searched.

The definitions of the strategies are given in the following table, along with the number of nodes searched for a simplified refinery scheduling problem involving about 2,000 constraints (500 nontrivially quadratic) and 1,500 variables. Due to the nature of the scheduling method, seven problems are sequentially solved in each run, each problem being a refinement of the previous problem. Solve time for the best strategy K was about 24 minutes on an 866 MHZ Pentium workstation with 256K RAM.

General conclusions are difficult because of the limited testing done, but our experience has been that those methods based on worst-case guarantees such as subdividing the variable with the largest gradient range do more poorly than adaptive methods which compute some measure of the actual deviation from linearity at the current trial solution.

Strategy F subdivided that variable which has the largest worst case gradient range (see equation (6)), so it deterministically reduces the worst-case maximum infeasibility (see (7)).

Strategy N is identical to Strategy F except that it only looks at those variables that are present in that equation that most strongly violates feasibility at the current trial solution.

Table 1. Results of subdivision strategy tests

Strategy	Description	Result
F	i: k_i=argmax(k_i:(G_{ki}-F_{ki}))	Failed, terminated after 6 hours
I	i: i=argmax(i:LPMI shadow price (z_i,w_i))	Failed, terminated after 6 hours
H	i: i=argmax(i:(v_i-u_i) *LPMI shadow price (z_i,w_i))	Success after 1269 nodes searched
A	j: j=argmax(j:|B_{kij}|(v_j-u_j)), k_i=argmax(k_i:(G_{ki}-F_{ki}))	Success after 837 nodes searched
J	i: i=argmax(i:z_i+w_i)	Success after 531 nodes searched
N	i: i=argmax(i:(G_{ki}-F_{ki})), k=argmax(k:$\Delta_k(x)$)	Success after 234 nodes searched
G	i: i=argmax(i:(G_{ki}-F_{ki})(v_i-u_i)), k=argmax(k:$\Delta_k(x)$)	Success after 197 nodes searched
B	j: j=argmax(j:|B_{kij}|(v_j-u_j)), i=argmax(i:(G_{ki}-F_{ki})), k=argmax(k:$\Delta_k(x)$)	Success after 123 nodes searched
K	i: i=argmax(i:z_i+w_i and x_i appears in quadratic term in constraint k), k=argmax(k:$\Delta_k(x)$)	Success after 89 nodes searched

Strategy A is a variation on Strategy F wherein it looks at the variable that contributes the most to the worst case gradient range.

Strategy B continues the series, it looks at the variable that contributes the most to the worst case gradient range but only among those variables that are present in that equation that most strongly violates feasibility at the current trial solution.

Strategy G was similar to Strategy B, except that it directly took that variable with the largest worst-case maximum infeasibility c.

The remaining strategies all use the enveloping LPMI problem (13) to help gauge which variable to subdivide.

Strategies I and H used the shadow prices given by the LPMI as a way of pricing the benefits of subdividing the variables, but this was relatively unrewarding.

Strategy J directly subdivides that variable which had the largest enveloping value, as measured by $z+w$.

Strategy K, which is currently our favorite, is a variation on Strategy J which subdivides the variable with the largest enveloping value within the most-infeasible equation.

3.5   Degree of Rigor

The current GCES algorithm was not designed to provide rigorous proofs of feasibility or infeasibility, but to provide a practical approach to the global feasibility of large systems of equations. It was designed to be cognizant of standard numerical error issues. Making it completely rigorous in the sense of Kearfott [10] would depend on resolution of the following issues.

Degree of feasibility of solutions in GCES are established by evaluating the equations at candidate solution points, measuring the resulting error in the equations, and accepting a solution within a desired epsilon. Error in the solution arises from errors in the e quations themselves, the numerical conditioning of the equations, and evaluation error, and can be made rigorous by adopting a rigorous arithmetic system such as Kearfott [10], as well as adding the specifications of bounds for the equations themselves.

Global infeasibility in GCES is established by a combination of bound propagation rules and the infeasibility of enveloping linear programs. The bound propagation rules used in GCES are numerically identical to the methods of interval arithmetic discussed in Kearfott [10] and so would be rigorous given rigorous bounds on the equations themselves.

The enveloping linear programs were also designed to be rigorous given rigorous bounds on the equations themselves. However, establishing infeasibility of the global equation then depends on the rigor with which the underlying linear program can be shown to be infeasible. GCES currently uses the CPLEX library, which is numerically robust but does not provide rigorous proofs of infeasibility. Providing rigor here depends on substituting a rigorous LP algorithm, similar to the suggestions made in Jansson [9].

3.6   Other Practical Aspects

As always, performance of the solver is critically dependent on the scaling of the variables. Not wishing to also take on the challenge of auto-scaling methods, the design choice was made in the current solver implementation to input typical magnitudes of the variables along with the bounding information, and to use those typical magnitudes for scaling.

We also looked at different strategies for subdividing the range of the chosen variable. The second-best approach was to divide the range into three pieces, taking one piece out of the interior of the range that contained the current linearization point. The best current strategy is to simply divide the range into five equal pieces.

The solver was coded in Java with interfaces to the CPLEX linear programming library and ran on standard 750MHz Pentium Windows workstations. The final test was on an instance of our refinery scheduling problem and had 13,711 variables with 17,892 constraints and equations, with 2,696 of the equations being nonlinear. The resulting system of equalities and inequalities was solved in 45 minutes.

3.7   Implementation of the Refinery Scheduler

The GCES solver has been implemented in Java code, with an interface to the CPLEX linear solver. The hybrid solver that uses the GCES solver is also implemented in Java, embedded in the refinery scheduler described in Section 2. Input data are stored in and drawn from Excel spreadsheets, with output in several formats, including Excel.

3.8   Next Steps

Benchmarking: All our tests to date have been against a particular large-scale refinery scheduling problem. In order to objectively evaluate this algorithm against other candidate algorithms, we would like to test it using standard test problems suites for global constraints.

Nonlinear Functions: Given the fact that we have a solver that can work with quadratic constraints, the current implementation can handle an arbitrary polynomial or rational function through rewriting and the introduction of additional variables. The issue is a heuristic one (system performance), not an expressive one. The GCES framework can be extended to include any nonlinear function for which one has analytic gradients, and for which one can compute reasonable function and gradient bounds given variable bounds.

Efficiency: While we have achieved several orders of magnitude speedup through the pragmatic measures described above (propagation, splitting strategy for sub-node generation, converging approximation), there is much yet to be done. In addition to further effort in the areas listed here, we intend to investigate the use of more sophisticated scaling techniques, and the adoption of generally-available nonlinear solvers for node solving.

Optimality: The current solver determines global feasibility, which is polynomial-equivalent to global optimality. We would like to add global optimality directly to the solver, first by replacing the current SLP feasibility subroutine by an NLP subroutine capable of finding local optimums within each feasible subregion, then by adding branch-and-bound to the overall subdivision search.

Scheduling Domains: In addition to refinery operations, we intend to extend our current implementation to hybrid systems which appear in other application domains, including batch manufacturing, satellite and spacecraft operations, and transportation and logistics planning.

Other Domains: The current hybrid solver is intended to solve scheduling problems. Other potential domains that we wish to investigate include abstract planning problems, the control of hybrid systems, and the simulation and solution of system behavior described as linear hybrid automata (LHA).

4   Summary

We have implemented a finite-capacity scheduler and an associated global equation solver capable of modeling and solving scheduling problems involving an entire petroleum refinery, from crude oil deliveries, through several stages of processing of intermediate material, to shipments of finished product. This scheduler employs an architecture described previously [4] for the coordinated operation of discrete and continuous solvers. There is considerable work remaining on all fronts. Nonetheless, the current solver has shown the ability to master problems previously found intractable.

References

[1]   Boddy, M., Carciofini, J., and Hadden, G.: Scheduling with Partial Orders and a Causal Model, Proceedings of the Space Applications and Research Workshop, Johnson Space Flight Center, August 1992

[2]   Boddy, M., and Goldman, R.: Empirical Results on Scheduling and Dynamic Backtracking, Proceedings of the International Symposium on Artificial Intelligence, Robotics, and Automation for Space, Pacadena, CA, 1994

[3]   Boddy, M., White, J., Goldman, R., and Short, N.: Integrated Planning and Scheduling for Earth Science Data Processing, Hostetter, Carl F., (Ed.), Proceedings of the 1995 Goddard Conference on Space Applications of Artificial Intelligence and Emerging Information Technologies, NASA Conference Publication 3296, 1995, 91-101

[4]   Boddy, M., and Krebsbach, K.: Hybrid Reasoning for Complex Systems, 1997 Fall Symposium on Model-directed Autonomous Systems

[5]   Cucker, F., and Smale, S.: Complexity Estimates Depending on Condition Number and Round-off Error, Jour. of the Assoc. of Computing Machinery, Vol 46 (1999) 113-184

[6]   Goldman, R., and Boddy, M.: Constraint-Based Scheduling for Batch Manufacturing, IEEE Expert, 1997

[7]   Heipcke, S.: Combined Modeling and Problem Solving in Mathematical Programming and Constraint Programming, PhD Thesis, School of Business, University of Buckingham, 1999

[8]   Hooker, J.: Logic-Based Methods for Optimization: Combining Optimization and Constraint Satisfaction, Wiley, John & Sons, 2000

[9]   Jansson, C.: Rigorous Error Bounds for the Optimal Value of Linear Programming Problems, Cocos '02, Global Constrained Optimization and Constraint Satisfaction, October 2 - 4, 2002

[10]   Kearfott, R.: Rigorous Global Search: Continuous Problems, Kluwer Academic Publishers, 1996

[11]   Lee, S., and Grossmann, I.: New Algorithms for Nonlinear Generalized Disjunctive Programming, Computers and Chem. Engng., 24(9-10), 2125-2141, 2000

[12] Moore, R.,: Methods and Applications of Interval Analysis, by Ramon E. Moore, Society for Industrial & Applied Mathematics, Philadelphia, 1979

[13] Smith, B., Brailsford, S., Hubbard, P., and Williams, H.: The Progressive Party Problem: Integer Linear Programming and Constraint Programming Compared, Working Notes of the Joint Workshop on Artificial Intelligence and Operations Research, Timberline, OR, 1995

[14] Van Hentenryck, P., McAllister, D., and Kapur, D.: Solving Polynomial Systems Using a Branch and Prune Approach, SIAM Journal on Numerical Analysis, 34(2), 1997

Consistency Techniques
for the Localization of a Satellite

Luc Jaulin

Laboratoire d'Ingénierie des Systèmes Automatisés
62 avenue Notre Dame du Lac, 49 000 Angers, France
jaulin@univ-angers.fr
http://www.istia.univ-angers.fr/~jaulin/

Abstract. This paper recalls that the problem of estimating the state vector of a nonlinear dynamic system can be interpreted as a constraint satisfaction problem over continuous domains with a large number (several thousands) of variables and constraints. Consistency techniques are then shown to be particularly efficient to contract the domains for the variables involved. This is probably due to the large number of redundancies naturally involved in the constraints of the problem. The approach is illustrated on the estimation of the position of a satellite in orbit around the earth.

1   Introduction

The problem of estimating the position of a satellite in orbit around the earth is a problem of state estimation of a nonlinear dynamic system (see Section 3). The aim of this paper is to show that basic consistency techniques working on continuous domains [3], [4], [8], [7], [19] can be particularly efficient to solve this problem.

Most of the existing methods for state estimation compute their solutions on a linear approximation of the dynamic system ([13], [17], see also [15] in the context of localization). These methods are very efficient. The results obtained are solutions of the linear approximation and are expected to be also an approximation of the solution of the initial nonlinear problem. This is usually false when strong nonlinearities are involved. Some interval methods (but without propagation) have also been proposed in the context of state estimation (see, e.g., [14]) but, to get an acceptable accuracy, a large number of bisections have to be performed, making the approach limited to low dimensional problems.

As linearization approaches, consistency techniques also compute an approximation of the solutions of the nonlinear problem. But their results are correct (*i.e.*, no solution is lost) and are not sensitive to strong nonlinearities or nondifferentiabilities in the dynamic system.

One of the main critic that can be done to consistency techniques is that they compute a superset of all solutions of the problem that may be huge even if the solution set is small. For state estimation, this pessimism is limited by presence of

C. Bliek et al. (Eds.): COCOS 2002, LNCS 2861, pp. 157–170, 2003.

a large number of redundant constraints and the treatment of difficult nonlinear state estimation problems can thus be considered. This can help to popularize consistency techniques. Note that some interesting results have already been obtained by consistency methods for the state estimation of a mobile robot in [12] or for the localization of a car in [2].

In Section 2 state estimation is shown to be a constraint satisfaction problem with many variables taking their values in the set of real numbers $\mathbb{R}$. In Section 3, the obtention of the state equations of the satellite is presented in details. A resolution of the localization problem is performed by the solver RE-ALPAVER [6] in Section 4. Section 5 presents some possible improvements to increase the efficiency of consistency techniques.

2   State Estimation is a Constraint Satisfaction Problem

A huge class of dynamic systems can be described, after discretization of the corresponding differential equation, by the following equations:

$$\begin{cases} \mathbf{x}(k+1) = \mathbf{f}(\mathbf{x}(k), \mathbf{w}(k), \mathbf{u}(k)) \\ \mathbf{y}(k) \quad = \mathbf{g}(\mathbf{x}(k)). \end{cases} \tag{1}$$

See Figure 1 for an illustration.

- $k \in [0, k_{\max}]$ is an integer which corresponds to discrete time. If t is the continuous time, we have the relation $t = k\delta$ where δ is the sampling time.
- $\mathbf{u}(k) \in \mathbb{R}^{n_u}$ is the input vector at time k. It can be chosen arbitrarily and is assumed to be known exactly.
- $\mathbf{y}(k) \in \mathbb{R}^{n_y}$ is the output vector at time k. It is measured with a given precision depending on the sensors used, i.e., for each components $y_i(k)$ of $\mathbf{y}(k)$ a bounding interval $[y_i](k)$ is available.
- $\mathbf{x}(k) \in \mathbb{R}^{n_x}$ is the state vector at time k. It can be assimilated to the memory of the system. It cannot be measured directly and nothing is known about it. State estimation aims at estimating its value, for all k.
- $\mathbf{w}(k) \in \mathbb{R}^{n_w}$ is the perturbation vector at time k. It is unknown but bounded. Its components can usually be assumed to belong to a small interval $[w_i](k)$ containing zero. This vector takes into account unmodeled dynamics of the actual plant (relativist effects, ...), unknown inputs (wind, ...) or an error due to the discretization procedure.
- The first equation $\mathbf{x}(k+1) = \mathbf{f}(\mathbf{x}(k), \mathbf{w}(k), \mathbf{u}(k))$ is called the *evolution equation* and describes the evolution of the system. Its form shows that the future of the system depends only of the state vector $\mathbf{x}(k)$ at the current time k and the present and future inputs $\mathbf{u}(k), \mathbf{u}(k+1), \ldots$ and perturbations $\mathbf{w}(k), \mathbf{w}(k+1), \ldots$ Thus, the vector $\mathbf{x}(k)$ summarizes everything that happened to the system in the past and that may influence the future behavior of the system.
- The second equation $\mathbf{y}(k) = \mathbf{g}(\mathbf{x}(k))$ is the *observation equation*. Each output $y_i(k)$ is a function the state vector at time k.

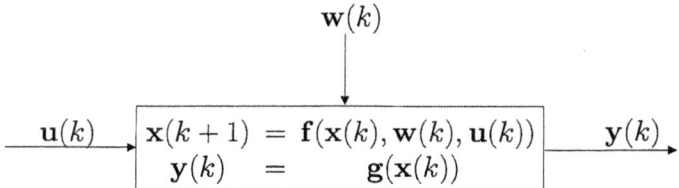

Fig. 1. Illustration of a discrete-time dynamic system

Remark 1. In practice, it is impossible to give a reliable measure of the error between the actual plant and the unperturbed model obtained by assuming that $\mathbf{w}(k) = 0$ in (1). Thus the domains $[w_i](k)$ chosen for the perturbation variables are not guaranteed to contain the true values for $w_i(k)$. The results obtained by a reliable method may thus be unreliable. When no solution exists for our state estimation problem, we are certain that too optimistic bounds have been chosen for $w_i(k)$. The associated domains should then be enlarged (see Section 5).

Our problem can be represented (see [12]) by a *constraint satisfaction problem* (CSP) where the set of $(k_{\max} + 2)\, n_x + (k_{\max} + 1)\,(n_w + n_y)$ variables is

$$
\begin{aligned}
\mathcal{V} = \{ & x_1(0), \ldots, x_{n_x}(0), x_1(1), \ldots, x_{n_x}(1), \ldots, \\
& \ldots, x_1(k_{\max} + 1), \ldots, x_{n_x}(k_{\max} + 1), \\
& w_1(0), \ldots, w_{n_w}(0), w_1(1), \ldots, w_{n_w}(1), \ldots, \\
& \ldots, w_1(k_{\max}), \ldots, w_{n_w}(k_{\max}), \\
& y_1(0), \ldots, y_{n_y}(0), y_1(1), \ldots, y_{n_y}(1), \ldots, \\
& \ldots, y_1(k_{\max}), \ldots, y_{n_y}(k_{\max}) \},
\end{aligned}
\tag{2}
$$

the set of domains is

$$
\begin{aligned}
\mathcal{D} = \{ & [x_1](0), \ldots, [x_{n_x}](0), [x_1](1), \ldots, [x_{n_x}](1), \ldots, \\
& \ldots, [x_1](k_{\max} + 1), \ldots, [x_{n_x}](k_{\max} + 1), \\
& [w_1](0), \ldots, [w_{n_w}](0), [w_1](1), \ldots, [w_{n_w}](1), \ldots, \\
& \ldots, [w_1](k_{\max}), \ldots, [w_{n_w}](k_{\max}), \\
& [y_1](0), \ldots, [y_{n_y}](0), [y_1](1), \ldots, [y_{n_y}](1), \ldots, \\
& \ldots, [y_1](k_{\max}), \ldots, [y_{n_y}](k_{\max}) \},
\end{aligned}
\tag{3}
$$

and the set of $(k_{\max} + 1)\,(n_x + n_y)$ constraints is

$$
\begin{aligned}
\mathcal{C} = \{ & x_1(1) = f_1\left(x_1(0), \ldots, x_{n_x}(0), w_1(0), \ldots, w_{n_w}(0), u_1(0), \ldots, u_{n_u}(0)\right), \\
& \ldots, \\
& x_{n_x}(1) = f_{n_x}\left(x_1(0), \ldots, x_{n_x}(0), w_1(0), \ldots, w_{n_w}(0), u_1(0), \ldots, u_{n_u}(0)\right), \\
& y_1(0) = g_1\left(x_1(0), \ldots, x_{n_x}(0)\right), \ldots, \\
& y_{n_y}(0) = g_{n_y}\left(x_1(0), \ldots, x_{n_x}(0)\right), \\
& \ \vdots \qquad\qquad\qquad \},
\end{aligned}
$$

where $f_1, \ldots, f_{n_x}$ and $g_1, \ldots, g_{n_y}$ are the coordinate functions associated with $\mathbf{f}$ and $\mathbf{g}$. Note that since the $u_i(k)$ are assumed to be exactly known, they have

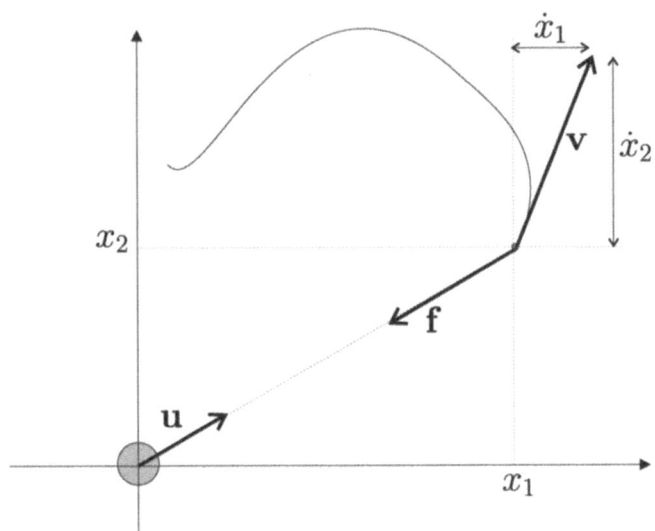

Fig. 2. Earth at coordinate $(0,0)$ and the satellite at coordinate (x_1, x_2)

not been considered as a variable of the CSP. In our formulation, it as been assumed that for all k, domains $[x_i](k)$, $[w_i](k)$ and $[y_i](k)$ containing the variables $x_i(k), w_i(k)$ and $y_i(k)$ are available. This is true for the measured outputs $y_i(k)$, this is also true for the bounded perturbations $w_i(k)$, but nothing is known about the values of the values of the state variable $x_i(k)$. Therefore, the corresponding domains should be taken as $[x_i](k) =]-\infty, \infty[$.

3   Formulation of Satellite Localization's Problem

As an application, consider the satellite, with coordinates (x_1, x_2) in orbit around the earth as illustrated on Figure 2.

Newton's universal law of gravitation states that the earth exerts a gravitational force $\mathbf{f}$ of attraction on the satellite. The direction of $\mathbf{f}$ is along the unit vector $\mathbf{u}$ directed by the line joining the two objects. The magnitude of the force is proportional to the product of the gravitational masses of the objects, and inversely proportional to the square of the distance between them. We have

$$\mathbf{f} = -G\frac{Mm}{x_1^2 + x_2^2}\mathbf{u}, \tag{4}$$

where M is the mass of the earth, m is the mass of the satellite and $G = 6.67x10^{-11}Nm^2kg^{-2}$ is the Newton's constant. Since

$$\mathbf{u} = \frac{1}{\sqrt{x_1^2 + x_2^2}}\begin{pmatrix} x_1 \\ x_2 \end{pmatrix}, \tag{5}$$

we have

$$\mathbf{f} = -G\frac{Mm}{\sqrt{x_1^2 + x_2^2}^3}\begin{pmatrix} x_1 \\ x_2 \end{pmatrix}. \tag{6}$$

Now, from the Newton's second law, we have

$$\mathbf{f} = m\frac{d\mathbf{v}}{dt} = m\begin{pmatrix} \ddot{x}_1 \\ \ddot{x}_2 \end{pmatrix}. \tag{7}$$

where $\mathbf{v}$ is the velocity of the satellite. Thus

$$\begin{pmatrix} \ddot{x}_1 \\ \ddot{x}_2 \end{pmatrix} = -G\frac{M}{\sqrt{x_1^2 + x_2^2}^3}\begin{pmatrix} x_1 \\ x_2 \end{pmatrix} \tag{8}$$

Set $\alpha = GM$, then

$$\begin{cases} \ddot{x}_1 = -\frac{\alpha x_1}{\sqrt{x_1^2 + x_2^2}^3} \\ \ddot{x}_2 = -\frac{\alpha x_2}{\sqrt{x_1^2 + x_2^2}^3} \end{cases} \tag{9}$$

For simplicity, we shall assume that $\alpha = 1$. To obtain a first order equation (to get state equations), we have to introduce two other variables x_3 and x_4 which satisfy

$$\begin{cases} \dot{x}_1 = x_3 \\ \dot{x}_2 = x_4 \end{cases} \tag{10}$$

Then we have the following continuous state equations:

$$\begin{cases} \dot{x}_1 = x_3 \\ \dot{x}_2 = x_4 \\ \dot{x}_3 = -\frac{x_2}{\sqrt{x_1^2 + x_2^2}^3} \\ \dot{x}_4 = -\frac{x_1}{\sqrt{x_1^2 + x_2^2}^3} \end{cases} \tag{11}$$

In practice, we are not able to measure the state directly. Instead, we shall assume that we are able to measure three quantities:

- y_1 : the angle between the direction satellite and a fixed direction (here, we took the tangent y_1 of the angle to get polynomial relations);
- y_2 : the radius speed of the satellite (which is equal to zero if the distance between the satellite and the earth remains constant). This can be done by the Doppler shift method which applies to electromagnetic waves in all portions of the spectrum. Radiation is blueshifted when the satellite moves away, and is redshifted when it comes near the earth;
- y_3 : the distance between the satellite and the earth.

The observation equations can thus be described by:

$$\begin{cases} y_1 = x_2/x_1 \\ y_2 = x_1 x_3 + x_2 x_4 \\ y_3 = x_1^2 + x_2^2 \end{cases} \tag{12}$$

In order to get discrete-time state equation, a discretization of the continuous state equation (11) should be performed. Using a first order approximation:

$$\mathbf{x}(t + \delta) \simeq \mathbf{x}(t) + \delta \mathbf{f}\left(x\left(t\right)\right), \tag{13}$$

where the sampling time is taken as $\delta = 0.5\,\text{sec}$, from (11) and (12), we get:

$$
\begin{cases}
x_1(k+1) = x_1(k) + 0.5\, x_3(k) + w_1(k) \\
x_2(k+1) = x_2(k) + 0.5\, x_4(k) + w_2(k) \\
x_3(k+1) = x_3(k) - 0.5\, \dfrac{x_1(k)}{\sqrt{x_1^2(k)+x_2^2(k)}^3} + w_3(k) \\
x_4(k+1) = x_4(k) - 0.5\, \dfrac{x_2(k)}{\sqrt{x_1^2(k)+x_2^2(k)}^3} + w_4(k) \\
y_1(k) \quad = x_2(k)/x_1(k) \\
y_2(k) \quad = x_1(k)x_3(k) + x_2(k)x_4(k) \\
y_3(k) \quad = x_1^2(k) + x_2^2(k)
\end{cases} \tag{14}
$$

Here, we have added perturbation variables w_1, w_2, w_3 and w_4 to take into account errors due to the discretization procedure and other unmodeled dynamics (presence of other planets, tide effects, ...). Note that it is impossible in practice to find rigorous bounds for w_i. For our example, we shall take $[w_i](k) = [-0.01, 0.01]$. For the output, the sensors are usually capable of giving rigorous intervals for the actual output $y_i(k)$. Here, we shall assume that the domains $[y_i(k)]$ for $y_i(k)$ is given by $[\check{y}_i(k) - 0.1, \check{y}_i(k) + 0.1]$, where $\check{y}_i(k)$ is the measure (*i.e.*, an approximation) of $y_i(k)$.

Again, the formulation of our system is a constraint satisfaction problem. For $k_{\max} = 200$, the set of 2215 variables is

$$
\begin{aligned}
\mathcal{V} = \{ & x_1(0), x_2(0), x_3(0), x_4(0), \ldots, x_1(201), x_2(201), x_3(201), x_4(201), \\
& w_1(0), w_2(0), w_3(0), w_4(0), \ldots, w_1(200), w_2(200), w_3(200), w_4(200), \quad (15) \\
& y_1(0), y_2(0), y_3(0), \ldots, y_1(200), y_2(200), y_3(200) \},
\end{aligned}
$$

the set of domains is

$$
\begin{aligned}
\mathcal{D} = \{ &]-\infty, \infty[,]-\infty, \infty[, \ldots,]-\infty, \infty[, \\
& [-0.01, 0.01], [-0.01, 0.01], \ldots, [-0.01, 0.01], \quad (16) \\
& [\check{y}_1(0) - 0.1, \check{y}_1(0) + 0.1], \ldots, [\check{y}_3(200) - 0.1, \check{y}_3(200) + 0.1] \}.
\end{aligned}
$$

The set of 1407 constraints of the CSP is given by the state equations given in formula (14).

Remark 2. The measurements $\check{y}_i(k)$ have been generated by simulation (with (14)) with the following initial conditions

$$\mathbf{x}(0) = (4,\ 0.5,\ -0.2,\ 0.5)^{\mathrm{T}}.$$

Note that other initial conditions could have lead to the same measures. This is the case for the initial vector

$$\mathbf{x}'(0) = (-4,\ -0.5,\ 0.2,\ -0.5)^{\mathrm{T}}.$$

The fact that different initial solutions can generate the same measurements (even in a noise-free context) is known as an identifiability problem [20]. Therefore, even without noise we will never be able to compute the initial conditions in a unique way. To avoid this identifiability problem we shall restrict the domains for $x_1(0), x_2(0), x_3(0)$ and $x_4(0)$ to $[0, \infty[, [0, \infty[,]-\infty, 0]$ and $[0, \infty[$, respectively.

Remark 3. Although the CSP described above contains 1407 constraints for 2215 variables, it can be considered as redundant. This is due to the fact that many of these variables (here exactly 1407 : the measurements $y_i(k)$ and the perturbation $w_i(k)$) have small domains and can thus be considered as approximately known.

4 Resolution

The contraction of the domains have been performed by the solver REALPAVER by using a basic local consistency method. Some lines of the program accepted by REALPAVER are given by:

```
Bisection none;
x1_0 in [0,+oo[,
x2_0 in [0,+oo[,
x3_0 in [-oo,0[,
x4_0 in [0,+oo[,
w1_0 in [-0.01,0.01],
w2_0 in [-0.01,0.01],
w3_0 in [-0.01,0.01],
w4_0 in [-0.01,0.01],
y1_0 in [0.107329,0.307329],
y2_0 in [-0.71049,-0.51049],
y3_0 in [16.117,16.317],
x1_1 in ]-oo,+oo[,
x2_1 in ]-oo,+oo[,
x3_1 in ]-oo,+oo[,
x4_1 in ]-oo,+oo[,

...

w1_200 in [-0.01,0.01],
w2_200 in [-0.01,0.01],
w3_200 in [-0.01,0.01],
w4_200 in [-0.01,0.01],
y1_200 in [-2.42001,-2.22001],
y2_200 in [-0.925105,-0.725105],
y3_200 in [136.612,136.812],
x1_201 in ]-oo,+oo[,
x2_201 in ]-oo,+oo[,
x3_201 in ]-oo,+oo[,
x4_201 in ]-oo,+oo[;
```

Constraints

```
x1_1=x1_0+0.5*x3_0+ w1_0,
x2_1=x2_0+0.5*x4_0+ w2_0,
x3_1=x3_0-0.5*x1_0/(sqrt(x1_0^2+x2_0^2))^3+w3_0,
x4_1=x4_0-0.5*x2_0/(sqrt(x1_0^2+x2_0^2))^3+w4_0,
y1_0=x2_0/x1_0,
y2_0=x1_0*x3_0+x2_0*x4_0,
y3_0=x1_0^2+x2_0^2,
...
x1_201=x1_200+0.5*x3_200+ w1_200,
x2_201=x2_200+0.5*x4_200+ w2_200,
x3_201=x3_200-0.5*x1_200/(sqrt(x1_200^2+x2_200^2))^3+w3_200,
x4_201=x4_200-0.5*x2_200/(sqrt(x1_200^2+x2_200^2))^3+w4_200,
y1_200=x2_200/x1_200,
y2_200=x1_200*x3_200+x2_200*x4_200,
y3_200=x1_200^2+x2_200^2;
```

The whole program can be downloaded at
http://www.istia.univ-angers.fr/~jaulin/satellite.txt. On a Pentium 300, REAL-PAVER returns the following lines

```
x1_0 in [3.8914121266466 , 4.017994154622222]
x2_0 in [0.4176613721408529 , 0.9567337119326976]
x3_0 in [-0.4113978603389778 , -0.1470621703041645]
x4_0 in [0.2009142731412085 , 0.9312204034312483]
y1_0 in [0.107329 , 0.2536019303090482]
y2_0 in [-0.7104900000000001 , -0.51049]
y3_0 in [16.117 , 16.317]
...
x1_200  in [4.385557786959884 , 4.885898743627584]
x2_200  in [-10.84674406984067 , -10.61309370002079]
x3_200  in [-0.1485890863687774 , +0.541771027979396]
x4_200  in [-0.0002353318428811776 , +0.3258643220373622]
y1_200  in [-2.42001 , -2.22001]
y2_200  in [-0.9251050000000001 , -0.725105]
y3_200  in [136.612 , 136.812]
...
END OF SOLVING
  Property:    reliable process (no solution is lost)
  Elapsed time: 810 ms
```

Figure 3 represents the actual values for the positions $(x_1(k), x_2(k))$ for the satellite and for $k \in [0, 200]$ that have been obtained by a simulation.

Figure 4 represents the boxes $[x_1](k) \times [x_2](k)$. This figure is consistent with the fact that for all k, the actual position vector $(x_1(k), x_2(k))$ obtained by the

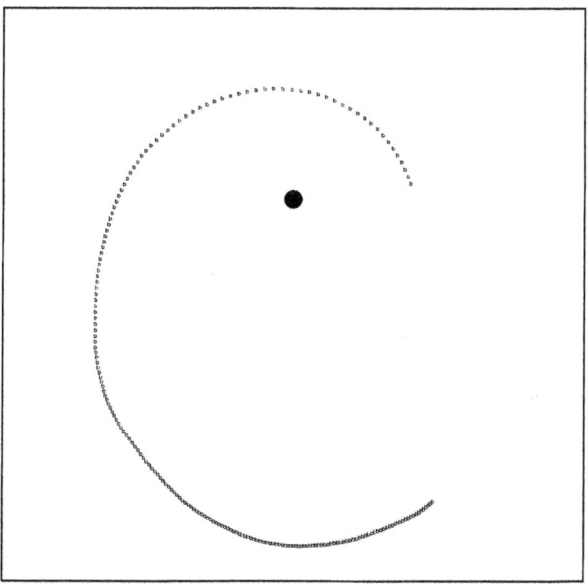

Fig. 3. Actual position of the satellite for k varying from 0 to 200

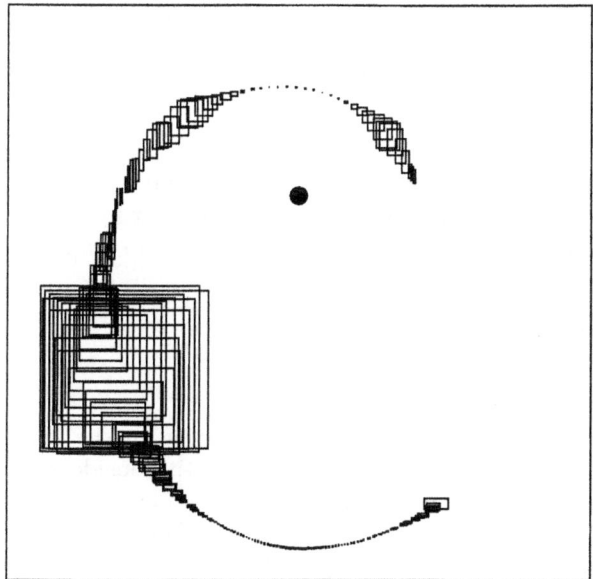

Fig. 4. Set of boxes generated by REALPAVER containing the actual positions for the satellite for different k in $[0, 200]$

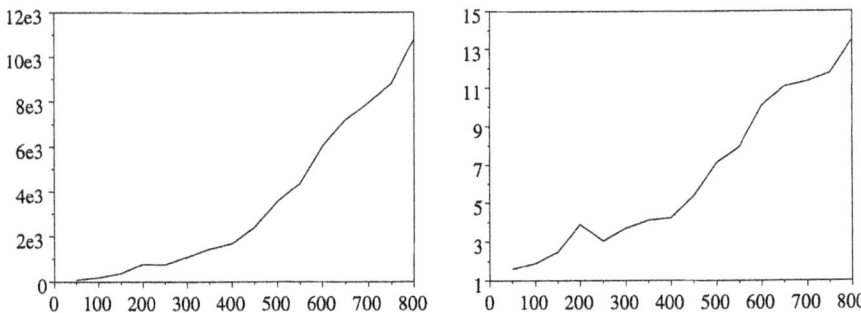

Fig. 5. Left: computing time t_c (in ms) with respect to $k_{\max}$; Right, $t_c/k_{\max}$ with respect to $k_{\max}$

simulation is an element of the box $[x_1](k) \times [x_2](k)$. For some k, the approximation obtained is too rough. This is probably due to the fact that in the associated regions for the satellite, the measures are not informative enough and that the local consistency used is not strong enough to get an accurate enclosure for the state vector.

Figure 5 (left) represents the computing time (t_c) of REALPAVER for different $k_{\max}$. The right part represents $t_c/k_{\max}$ with respect to $k_{\max}$. These experimental results suggest that the complexity of the local consistency algorithm used by REALPAVER could be quadratic with respect the number of variables. Note that when $k_{\max} = 800$, the CSP treated by REALPAVER is huge since it contains 8815 variables for 5607 constraints. To my knowledge, such a huge CSP related to an application has never been treated before.

Remark 4. The constraint graph has here a rather uncommon structure (studied in [11]) which is not taken into account by REALPAVER. A more efficient approach would be to alternate a forward and a backward propagation until a fixed point is reached. This approach has been presented in [10].

5 Possible Improvements

5.1 Using Linear Consistencies

The contraction method used for the localization of the satellite is based on hull consistency. Now, hull consistency is local (it considers the constraints independently) and pessimism exists when a variable occurs more than once in a constraint (due to the dependency problem). This pessimism can be reduced by using box-consistency [1]. More global (i.e., taking the dependency between two or more constraints into account) consistency approaches can also be considered [16], [18], but most of them are very sensitive to the number of variables which is very high for state estimation problems.

Now, for most state estimation problems, it is always possible to find large subnetworks of the associated problem which involve only linear constraints. The domains for the variables of the linear subnetworks can then be contracted by linear techniques and propagated back to the original CSP. This is illustrated by the following example

Example 1. Consider the example of a polynomial CSP, inspired by our satellite localization problem, with 7 variables and 3 constraints given by

$$
C : \begin{cases} x_1' = x_1 + x_3 \\ x_2' = x_4 \\ x_3' = x_3 - \dfrac{x_1}{\sqrt{x_1^2 + x_2^2}^3} \end{cases} \tag{17}
$$

This set of constraints can be decomposed into the 8 following primitive constraints:

$$
C_p : \begin{cases} x_1' = x_1 + x_3 \\ x_2' = x_4 \\ x_3' = x_3 - z_5 \\ z_1 = x_1^2 \\ z_2 = x_2^2 \\ z_3 = z_1 + z_2 \\ z_4 = z_3^{-3/2} \\ z_5 = x_1 z_4 \end{cases} \tag{18}
$$

Assume that all domains of the nonlinear variables of C_p are included in $]0, \infty[$. If this is not the case, the constraint involving a variable that may be negative should be removed. Then C_p can be decomposed as follows:

$$
C_1 : \begin{cases} x_1' = x_1 + x_3 \\ x_2' = x_4 \\ x_3' = x_3 - z_5 \\ z_3 = z_1 + z_2 \end{cases} \quad C_2 : \begin{cases} \hat{z}_1 = \log(z_1) \\ \hat{z}_2 = \log(z_2) \\ \hat{z}_4 = \log(z_3) \\ \hat{z}_5 = \log(z_4) \\ \hat{x}_1 = \log(x_1) \\ \hat{x}_2 = \log(x_2) \end{cases} \quad C_3 : \begin{cases} \hat{z}_1 = 2\hat{x}_1 \\ \hat{z}_2 = 2\hat{x}_2 \\ \hat{z}_4 = -\frac{3}{2}\hat{z}_3 \\ \hat{z}_5 = \hat{x}_1 + \hat{z}_4 \end{cases} \tag{19}
$$

A linear solver can thus be used to contract optimally C_1 and C_3. A local consistency technique is here optimal for C_2. The resulting contraction can then be used to contract the original CSP (17).

5.2 Tuning the Initial Size for the Domains

In practice, an efficient contraction method is needed but global consistency is not always desirable. Indeed, our model is only an approximation of the reality and ignores many unknown effects. As a result, the solution set of the CSP is often empty and global consistency would result in a loss of any information we had about the position of the satellite. A possible way to keep a satisfactory result is to use a parameter β to tune the size of the perturbation and output intervals. This parameter is decreased until the envelope generated by our favorite contraction method is accurate enough. For our satellite problem, the following algorithm can be chosen.

1. $\beta = \beta_0$, $\ell = 1$;
2. for $i \in \{1, 2, 3, 4\}$, $k \in \{0, \ldots, 200\}$, $[w_i(k)] = [-0.01\beta, 0.01\beta]$;
3. for $i \in \{1, 2, 3\}$, $k \in \{0, \ldots, 200\}$, $[y_i(k)] = [\check{y}_i(k) - 0.1\beta, \check{y}_i(k) + 0.1\beta]$;
4. $\mathcal{S}(\ell) =$ approximation of the solution set of the CSP obtained by the available consistency method;
5. if $\mathcal{S}(\ell) \neq \emptyset$, $\ell = \ell + 1$; decrease β, go to 2;
6. else return $\mathcal{S}(\ell - 1)$; END;

In this algorithm, β_0 is large enough to make to CSP satisfiable at the first step. This algorithm

— never returns the empty set. This is important for the engineer, who often needs a punctual estimation of for state vectors in order to take his decision,
— can tackle nonlinear constraints much more easily than most existing methods which are not based on consistency algorithm,
— can produce a result with an accurate approximation in a short time, depending on the consistency used.

Of course, the algorithm is not anymore correct but the correctness is not as fundamental for engineer: since many unknown effects (relativity, presence of other planets, shape of the earth, ...) are always neglected in the equations, the model is never correct. However, I believe that the above algorithm may often produce *more correct* results than most existing nonlinear approaches which are generally based on the linearization of the equations of the dynamic system.

6   Conclusion

In this paper, consistency techniques have been used to deal with the difficult problem of nonlinear state estimation which has been recalled to be a constraint satisfaction problem with continuous variables. The large number of measurements that can generally be performed on the dynamic system introduces many redundancies which benefit to constraint propagation methods. Moreover, since in practice bounds of some variables can be violated and since many other unknown effects of the dynamic system are neglected, the solution set of the CSP to be solved is generally empty. The pessimism of consistency methods may thus counterbalance the fact that no solution exists. An elementary algorithm to tune the balance parameter β has been proposed to give satisfaction to engineers who need a reasonably small (but nonempty) envelope for the state vectors to the detriment of the correctness of the results.

The approach presented here has supposed that the dynamic system was described by a discrete-time state equation. Now, in practice, the system is described by nonlinear differential equations. The discretization procedure adds some errors that could be avoided. An efficient approach working directly on the differential equation could be obtained by combining the approach presented in [5] with that in [9].

References

[1] Benhamou, F., Goualard, F., Granvilliers, L., Puget, J. F.: Revising Hull and Box Consistency. Proceedings of the International Conference on Logic Programming, Las Cruces, NM, (1999) 230–244

[2] Bouron, R.: Méthodes ensemblistes pour le diagnostic, l'estimation d'état et la fusion de données temporelles. Université de Compiègne, PhD dissertation, (2002)

[3] Cleary, J. G.: Future Computing Systems. Logical arithmetic, 2(2), (1987) 125–149

[4] E. Davis, E.: Constraint propagation with interval labels. Artificial Intelligence, 32(3) (1987) 281–331

[5] Deville, Y., Janssen, M., Van Hentenryck, P.: Consistency techniques in ordinary differential equations. Proceedings of the Fourth International Conference on Principles and Practice of Constraint Programming. Lecture Notes in Computer Science, Springer Verlag, Berlin Heidelberg New York (1998)

[6] Granvilliers, L.: RealPaver, available at http://www.sciences.univ-nantes.fr /info/perso/permanents/granvil/realpaver/. IRIN, University of Nantes (2002)

[7] van Hentenryck, P., Deville, Y., Michel, L.: Numerica: A Modeling Language for Global Optimization. MIT Press, Boston, MA, (1997)

[8] Hyvönen, E.: Constraint reasoning based on interval arithmetic; The tolerance propagation approach. Artificial Intelligence, 58(1-3), (1992) 71-112

[9] Jaulin, L.: Nonlinear bounded-error state estimation of continuous-time systems. Automatica. 38 (2002) 1079–1082

[10] Jaulin, L., Braems, I., Kieffer, M., Walter, E.: Nonlinear State Estimation Using Forward-Backward Propagation of Intervals. W. Kraemer and J. W. Gudenberg (Eds), Scientific Computing, Validated Numerics, Interval Methods, Proceedings of SCAN 2000, Kluwer Academic Publishers (2001) 191-204

[11] Jaulin, L., Kieffer, M., Braems, I., Walter E.: Guaranteed Nonlinear Estimation Using Constraint Propagation on Sets. International Journal of Control, 74 (18), (2001) 1772–1782

[12] Jaulin, L., Kieffer, M., Didrit, O., Walter, E.: Applied Interval Analysis, with Examples in Parameter and State Estimation, Robust Control and Robotics, Springer-Verlag, London, (2001)

[13] Kalman, R. E.: A New Approach to Linear Filtering and Prediction Problems. Transactions of the AMSE, Part D, Journal of Basic Engineering, 82 (1960) 35–45

[14] Kieffer, M., Jaulin, L., Walter, E., Meizel, D.: Robust Autonomous Robot Localization Using Interval Analysis. Reliable Computing, 6(3), (2000) 337–362

[15] Leonard, J. J., Durrant-Whyte, H. F.: Mobile robot localization by tracking geometric beacons, IEEE Transactions on Robotics and Automation, 7(3), (1991) 376–382

[16] Lhomme, O.: Consistency Techniques for Numeric CSPs. Proceedings of the International Joint Conference on Artificial Intelligence, Chambéry, France, (1993) 232–238

[17] Ljung, L.: Asymptotic Behavior of the Extended Kalman Filter as a Parameter Estimator for Linear Systems. IEEE Transactions on Automatic Systems, **24**, (1979) 36–50

[18] Sam-Haroud, D. J., Faltings, B.: Consistency Techniques for Continuous Constraints. Constraints, **1**(1-2), (1996) 85–118

[19] VanEmden, M.: Algorithmic Power from Declarative Use of Redundant Constraints. Constraints, **4**(4), (1999) 363–381

[20] Walter, E.,Identifiability of state space models, Springer-Verlag, Berlin, Germany, (1982)

Computing Interval Parameter Bounds from Fallible Measurements Using Overdetermined (Tall) Systems of Nonlinear Equations*

G. William Walster[1] and Eldon R. Hansen[2]

[1] Sun Microsystems Laboratories
[2] Consultant

Abstract. Overdetermined (tall) systems of nonlinear equations naturally arise in the context of computing interval parameter bounds from fallible data. In tall systems, there are more interval equations than unknowns. As a result, these systems can appear to be inconsistent when they are not. An algorithm is given to compute interval nonlinear parameter bounds from fallible data and to possibly prove that no bounds exist because the tall system is inconsistent.

1 Parameter Estimation in Nonlinear Models

In [3] Lang used interval arithmetic to perform the analysis of fallible observations from an experiment to compute bounds on Newton's constant of gravitation G. Because the computed bounds were sufficiently different from the then accepted approximate value, subsequent experiments were conducted to refine the accepted approximate value and the interval bound on G. See [5].

Using the traditionally accepted methodology to compute approximate parameter values from nonlinear models of observable data requires a number of questionable assumptions. In the best case, if all assumptions are satisfied, the final result is a less than 100% statistical confidence interval rather than a containing interval bound. For example, the method of least squares produces a solution approximation even when the data on which it is based are inconsistent.

A better procedure is to solve a system of interval nonlinear equations using the interval version of Newton's method. If assumptions for this procedure are satisfied, the result is a guaranteed bound on the parameter(s) in question. If assumptions are sufficiently violated and enough observations are available, the procedure can *prove* the system of equations and interval data are inconsistent. This better procedure is now described.

C. Bliek et al. (Eds.): COCOS 2002, LNCS 2861, pp. 171–177, 2003.
© Springer-Verlag Berlin Heidelberg 2003

1.1   Nonlinear Parameter Estimation

Let n q-element vector measurements, $\mathbf{x}_1, \cdots, \mathbf{x}_n$ with $\mathbf{x}_i = (x_{i1}, \cdots, x_{iq})^T$ be given. Assume these measurements depend on the value of a p-element vector parameter, $\boldsymbol{\theta} = (\theta_1, \cdots, \theta_p)^T$. Moreover, assume an analytic model exists for the observation vectors, $\mathbf{x}_i$, as a function of $\boldsymbol{\theta}$ and the true value $\mathbf{c}_i = (c_{i1}, \cdots, c_{ir})^T$ of conditions under which the $\mathbf{x}_i$ are measured. Thus:

$$\mathbf{x}_i = \mathbf{h}(\boldsymbol{\theta} \mid \mathbf{c}_i). \tag{1}$$

The problem is to construct interval bounds $\boldsymbol{\theta}^{\mathbf{I}}$ on the elements of $\boldsymbol{\theta}$ from interval bounds $\mathbf{x}^{\mathbf{I}}_i$ on the fallible measurements, $\mathbf{x}_i$ and interval bounds $\mathbf{c}^{\mathbf{I}}_i$ on the conditions of measurement.

1.2   Interval Observation Bounds

The development of interval measurement bounds begins by recognizing that a measurement x can be modeled (or thought of) as an unknown value t to which an error is added from the interval

$$\varepsilon \times [-1, 1] = \varepsilon^{\mathbf{I}}, \tag{2}$$

where $0 \leq \varepsilon$. No assumption is made about the distribution of individual measurement errors from the interval $\varepsilon^{\mathbf{I}}$ that are added to t in the process of measuring x.

At once it follows that

$$x \in t + \varepsilon^{\mathbf{I}}. \tag{3}$$

More importantly, if the interval observation X is defined to be $x + \varepsilon^{\mathbf{I}}$, then

$$t \in X. \tag{4}$$

Enclosure (4) is an immediate consequence of the fact that zero is the midpoint of $\varepsilon^{\mathbf{I}}$. The implications of this simple idea were considered by Walster [6]. They are:

- Given multiple interval observations X_i, all of which are enclosures of t, so must their intersection. Therefore

$$t \in \bigcap_{i=1}^{n} X_i.$$

- Given random finite intervals X_i, all of which contain the value t, the expected width of their intersection decreases as n increases.
- An empty intersection is proof that:

$$t \notin X_i$$

for some value of i. This can be true either because

- the width of the interval observations X_i is too narrow,
- there is no single value t that is contained in all the interval measurements X_i, or
- both of the above.

The first alternative means that the assumption regarding the accuracy of the measurement process is false. The second alternative means that the model for the single common value t is false.

Walster [6] also explored how this simple idea works in practice to compute an interval bound on a common value under various probability distributions for values of the random variable $\varepsilon \in \varepsilon^{\mathbf{I}}$. He also discussed how this estimation principle can be generalized and used to bound parameters of nonlinear models given bounded interval observations, or observation vectors. The following is a more complete elaboration of the nonlinear generalization.

1.3   General Development

Given exact values of a set of conditions $\mathbf{c}_i = (c_{i1}, \cdots, c_{ir})^T$ under which observations $\mathbf{x}_i = (x_{i1}, \cdots, x_{iq})^T$ are made, assume the observation vectors, $\mathbf{x}_i$ ($i = 1, \cdots, n$), satisfy the following model:

$$\mathbf{x}_i \in \mathbf{h}(\boldsymbol{\theta} \mid \mathbf{c}_i) + \varepsilon(\boldsymbol{\theta}, \mathbf{c}_i) \times [-1, 1]; \tag{5}$$

where the vectors $\mathbf{0} \leq \varepsilon(\boldsymbol{\theta}, \mathbf{c}_i)$ bound unknown modeling and direct measurement errors. Specifically, if the p elements of $\boldsymbol{\theta}$ and the rn elements of all the $\mathbf{c}_i$ were known (in practice they are not), assume it would be possible to compute intervals

$$\varepsilon^{\mathbf{I}}(\boldsymbol{\theta}, \mathbf{c}_i) = \varepsilon(\boldsymbol{\theta}, \mathbf{c}_i) \times [-1, 1] \tag{6}$$

from which it follows immediately that

$$\mathbf{0} \in \mathbf{x}_i - \mathbf{h}(\boldsymbol{\theta} \mid \mathbf{c}_i) + \varepsilon^{\mathbf{I}}(\boldsymbol{\theta}, \mathbf{c}_i). \tag{7}$$

Note that (6) is a generalization of (2), and (7) is a generalization of (4) if written in the form

$$0 \in X - t.$$

If (as is normally the case) the conditions $\mathbf{c}_i$ under which measurements $\mathbf{x}_i$ are made are not known, but *are* contained in intervals $\mathbf{c}^{\mathbf{I}}_i$, then taking all bounded modeling and observation errors into account:

$$\mathbf{0} \in \mathbf{x}_i - \mathbf{h}(\boldsymbol{\theta} \mid \mathbf{c}^{\mathbf{I}}_i) + \varepsilon^{\mathbf{I}}(\boldsymbol{\theta}, \mathbf{c}^{\mathbf{I}}_i). \tag{8}$$

In this general form, the widths of interval measurements $\mathbf{x}^{\mathbf{I}}_i$ are themselves functions of both the unknown parameters $\boldsymbol{\theta}$ and fallibly measured conditions under which the measurements are made. That is:

$$\mathbf{x}^{\mathbf{I}}_i = \mathbf{x}_i + \varepsilon^{\mathbf{I}}(\boldsymbol{\theta}, \mathbf{c}^{\mathbf{I}}_i). \tag{9}$$

This interval observation model is the generalization of

$$X = x + \varepsilon^{\mathbf{I}}$$

which is consistent with (3). The interval observation model (9) is needed to solve for interval bounds on the parameter vector $\boldsymbol{\theta}$. If there is no solution for a given set of interval observation vectors $\mathbf{x}^{\mathbf{I}}{}_i$ and interval bounds on measurement conditions $\mathbf{c}^{\mathbf{I}}{}_i$, then either:

- the observation model $\mathbf{h}(\boldsymbol{\theta} \mid \mathbf{c}^{\mathbf{I}}{}_i)$ is false,
- the measurement error model $\varepsilon^{\mathbf{I}}(\boldsymbol{\theta}, \mathbf{c}^{\mathbf{I}}{}_i)$ is false,
- the computational system used to compute interval bounds on the elements of $\boldsymbol{\theta}$ is flawed, or
- some combination of the above.

In this way, and by eliminating alternative explanations, the theory represented in $\mathbf{h}(\boldsymbol{\theta} \mid \mathbf{c}^{\mathbf{I}}{}_i)$ or the observation error model represented in $\varepsilon^{\mathbf{I}}(\boldsymbol{\theta}, \mathbf{c}^{\mathbf{I}}{}_i)$ can be *proved* to be false.

1.4   The System of Nonlinear Equations

To guarantee any computed interval $\boldsymbol{\theta}^{\mathbf{I}}$ is indeed a valid bound on the true value of $\boldsymbol{\theta}$, the following must be true:

- the given model $\mathbf{h}$;
- the interval bounds $\mathbf{c}^{\mathbf{I}}{}_i$ on the conditions under which fallible measurements $\mathbf{x}_i$ are made; and,
- the model for interval bounds $\varepsilon^{\mathbf{I}}(\boldsymbol{\theta}, \mathbf{c}^{\mathbf{I}}{}_i)$ on observation errors.

To be consistent with the given models, all the actual measurement vectors $\mathbf{x}_i$ must satisfy relation (8). A logically equivalent, but more suggestive way to write this system of constraints is:

$$\mathbf{x}_i - \mathbf{h}(\boldsymbol{\theta} \mid \mathbf{c}^{\mathbf{I}}{}_i) + \varepsilon^{\mathbf{I}}(\boldsymbol{\theta}, \mathbf{c}^{\mathbf{I}}{}_i) = \mathbf{0} \quad (i = 1, \cdots, n). \tag{10}$$

When used in (10), a *possible* value of $\boldsymbol{\theta}$ produces intervals that contain zero for all i. Any value of $\boldsymbol{\theta}$ that fails to do this cannot be in the solution set of (10). Thus, (10) is just an interval system of nonlinear equations in the unknown parameter vector $\boldsymbol{\theta}$. Writing (10) as an equation is somewhat misleading, but is standard practice in interval analysis. The solution set is the set of vectors $\boldsymbol{\theta}$ for which a solution exists for some $\mathbf{c}_i \in \mathbf{c}^{\mathbf{I}}{}_i$ and some $\varepsilon \in \varepsilon^{\mathbf{I}}$. The difficulty is that the total number of scalar equations nq might be much larger than the number p of scalar unknowns in the parameter vector $\boldsymbol{\theta}$. Point (rather than interval) systems of equations where $p < nq$ are called "overdetermined". For interval nonlinear equations, this is a misnomer because the interval equations might or might not be consistent. As mentioned above, inconsistency (an empty solution set) is an informative event.

2   Solving Nonlinear Equations

Let $\mathbf{f} : \mathbb{R}^n \to \mathbb{R}^m$ $(n \leq m)$ be a continuously differentiable function. The parameter estimation problem in Section 1.4 is just a special case of the more general problem now considered. Table 1 shows the correspondence between the parameter estimation problem and equivalent nonlinear equations to be solved. Both unknowns and equations are shown. Having established this correspondence, the problem becomes to find and bound all the solution vectors of $\mathbf{f}(\mathbf{x}) = \mathbf{0}$ in a given initial box $\mathbf{x}^{\mathbf{I}(0)}$. For noninterval methods, it can sometimes be difficult to find reasonable bounds on a single solution, quite difficult to find reasonable bounds on all solutions, and generally impossible to know whether reasonable bounds on all solutions have been found. In contrast, it is a straightforward problem to find reasonable bounds on solutions in $\mathbf{x}^{\mathbf{I}(0)}$ using interval methods; and it is trivially easy to computationally determine that all solutions in $\mathbf{x}^{\mathbf{I}(0)}$ have been bounded. What is unusual in this problem is that the order m of $\mathbf{f}$ can be greater than the order p of $\mathbf{x}^{\mathbf{I}}$. A factor that simplifies obtaining solution(s) is the assumption that the equations are consistent. This has the effect of reducing the number of equations to the number of variables.

3   Linearization

Let $\mathbf{x}$ and $\mathbf{y}$ be points in a box $\mathbf{x}^{\mathbf{I}}$. Suppose we expand each component f_i $(i = 1, \cdots, m)$ of $\mathbf{f}$ by one of the procedures commonly used to linearize nonlinear equations to be solved using the interval Newton method. Define the matrix of partial derivatives of the elements f_i of $\mathbf{f}$ with respect to the elements x_j $(j = 1, \cdots, p)$ of $\mathbf{x}$:

$$\mathbf{J}_{ij} = \left(\frac{\partial f_i(\mathbf{x})}{\partial x_j} \right).$$

If $n = m$, the system is square, and $\mathbf{J}$ is the Jacobian of $\mathbf{f}$. This is the usual situation in which the interval Newton method is applied. See [2].

In passing it is worth noting that in place of partial derivatives, slopes can be used to good advantage. Slopes have narrower width than interval bounds

Table 1. Correspondence between Parameter Estimation and Nonlinear Equations

	Parameter Estimation	Nonlinear Equations
Unknowns	$\theta^{\mathbf{I}} = (\theta_1^{\mathbf{I}}, \cdots, \theta_p^{\mathbf{I}})^T$	$\mathbf{x}^{\mathbf{I}} = (X_1, \cdots, X_p)^T$
Equations	$\mathbf{x}_i - \mathbf{h}(\theta \mid \mathbf{c}^{\mathbf{I}}_i) + \varepsilon^{\mathbf{I}}(\theta, \mathbf{c}^{\mathbf{I}}_i) = 0$ $(i = 1, \cdots, n)$	$\mathbf{f} = (f_1, \cdots, f_m)^T = 0$ where $m = nq$

on derivatives and might exist when derivatives are undefined. Nevertheless, the remaining development uses derivatives as they are more familiar than slopes.

Combining the results in vector form:

$$\mathbf{f}(\mathbf{y}) \in \mathbf{f}(\mathbf{x}) + \mathbf{J}(\mathbf{x}, \mathbf{x}^{\mathbf{I}})(\mathbf{y} - \mathbf{x}). \qquad (11)$$

Even in the non-square situation, $\mathbf{J}$ is still referred to herein as the Jacobian of $\mathbf{f}$. The notation $\mathbf{J}(\mathbf{x}, \mathbf{x}^{\mathbf{I}})$ is used to emphasize the fact that both noninterval values $\mathbf{x}$ and intervals $\mathbf{x}^{\mathbf{I}}$ are used in practice to compute Jacobian matrix elements. See [2].

If $\mathbf{y}$ is a zero of $\mathbf{f}$, then $\mathbf{f}(\mathbf{y}) = \mathbf{0}$, and (11) is replaced by

$$\mathbf{f}(\mathbf{x}) + \mathbf{J}(\mathbf{x}, \mathbf{x}^{\mathbf{I}})(\mathbf{y} - \mathbf{x}) = \mathbf{0}. \qquad (12)$$

Define the solution set of (12) to be

$$\mathbf{s} = \left\{ \mathbf{y} \,\middle|\, \begin{array}{l} f_i(\mathbf{x}) + [\mathbf{J}(\mathbf{x}, \mathbf{x}'(i))(\mathbf{y} - \mathbf{x})]_i = 0, \\ \mathbf{x}'(i) \in \mathbf{x}^{\mathbf{I}}, \quad (i = 1, \cdots, n) \end{array} \right\}.$$

Note: $\mathbf{x}'(i)$ is a different (real) vector for each value of $i = 1, \cdots, n$. The set $\mathbf{s}$ contains any point $\mathbf{y} \in \mathbf{x}^{\mathbf{I}}$ for which $\mathbf{f}(\mathbf{y}) = \mathbf{0}$.

The smaller the box $\mathbf{x}^{\mathbf{I}}$, the smaller the set $\mathbf{s}$. The object of an interval Newton method is to reduce $\mathbf{x}^{\mathbf{I}}$ until $\mathbf{s}$ is as small as desired so that a solution point $\mathbf{y} \in \mathbf{x}^{\mathbf{I}}$ is tightly bounded. Note that $\mathbf{s}$ is generally not a box.

Normally, the system of linear equations in (12) is solved using any of a variety of interval methods. In the present situation if $n < m$, the linear system is overdetermined and therefore appears to be inconsistent. This is not necessarily the case. If the procedure described by Hansen and Walster [1] is used to compute an interval bound $\mathbf{y}^{\mathbf{I}}$ on $\mathbf{y}$, then $\mathbf{y}^{\mathbf{I}}$ contains the set of consistent solutions $\mathbf{s}$.

For the solution of (12), the standard and distinctive notation $N(\mathbf{x}, \mathbf{x}^{\mathbf{I}})$ is used in place of $\mathbf{y}.^{\mathbf{I}}$ This emphasizes the solution's dependence on both $\mathbf{x}$ and $\mathbf{x}^{\mathbf{I}}$. The leter N stands for Newton. However, the solution $N(\mathbf{x}, \mathbf{x}^{\mathbf{I}})$ can be obtained using algorithms other than Newton's method.

From (12), define an iterative algorithm of the form

$$\mathbf{f}(\mathbf{x}) + \mathbf{J}(\mathbf{x}, \mathbf{x}^{\mathbf{I}})(N(\mathbf{x}^{(k)}, \mathbf{x}^{\mathbf{I}(k)}) - \mathbf{x}) = \mathbf{0} \qquad (13a)$$

$$\mathbf{x}^{\mathbf{I}(k+1)} = \mathbf{x}^{\mathbf{I}(k)} \cap N(\mathbf{x}^{(k)}, \mathbf{x}^{\mathbf{I}(k)}) \qquad (13b)$$

for $k = 0, 1, 2, \cdots$ where $\mathbf{x}^{(k)}$ must be in $\mathbf{x}^{\mathbf{I}(k)}$. A good choice for $\mathbf{x}^{(k)}$ is the center $m(\mathbf{x}^{\mathbf{I}(k)})$ of $\mathbf{x}^{\mathbf{I}(k)}$. For details on computing $N(\mathbf{x}^{(k)}, \mathbf{x}^{\mathbf{I}(k)})$ when the system of interval linear equations appears to be overdetermined, see [1]. Solving the linear system in (12) with the algorithm described in [1] uses preconditioning followed by Gaussian elimination.

4 Conclusion

Computing bounds on nonlinear parameters from fallible observations is a pervasive problem. In the presence of uncertain observations, attempting to capture uncertainty with Gaussian error distributions is problematic when nonlinear functions of observations are computed. See for example, [4].

The procedure described in this paper uses the interval solution of a system of nonlinear equations to compute bounds on nonlinear parameters from fallible data. Among the many advantages of this approach is the ability to aggregate data from independent experiments, thereby continuously narrowing interval bounds. Whenever different interval results are inconsistent, or if the set of interval bounds from a given data set is empty, this *proves* an assumption is violated or the model for the observations is measurably wrong.

Narrower parameter bounds can be computed from a calibrated system. It is interesting to note that the same procedure as described above can be used to solve the calibration problem. All that must be done is to modify (8) in the following way:

- replace selected unknown parameter values with their now measured bounds $\theta_i^{\mathbf{I}}$, and
- solve for narrower bounds on any parameters in the model for $\varepsilon^{\mathbf{I}}\left(\boldsymbol{\theta}, \mathbf{c}^{\mathbf{I}}_i\right)$.

References

[1] E. Hansen and G. W. Walster. Solving overdetermined systems of interval linear equations. *Reliable Computing,* 2002. Submitted.

[2] E. R. Hansen. *Global Optimization Using Interval Analysis.* Marcel Dekker, Inc., New York, 1992.

[3] B. Lang. Verified quadrature in determining Newton's constant of gravitation. *Journal of Universal Computer Science,* 4(1):16-24, 1998. http://www.jucs.org/ jucs_4_1.

[4] National Institute of Standards and Technology (NIST). Fundamental physical constants, 1998.
http://physics.nist.gov/cuu/Constants/bibliography.html.

[5] The Eöt-Wash Group: Laboratory Tests of Gravitational Physics. The controversy over Newton's gravitational constant. http://www.npl.washington.edu/ eotwash/gconst.html.

[6] G. W. Walster. Philosophy and practicalities of interval arithmetic. In R. E. Moore, editor, *Reliability in Computing,* pages 309-323. Academic Press, Inc., San Diego, California, 1988.

Maintaining Global Hull Consistency
with Local Search for Continuous CSPs

Jorge Cruz and Pedro Barahona

Dep. de Informática, Universidade Nova de Lisboa
2829-516 Caparica, Portugal
{jc,pb}@di.fct.unl.pt

Abstract. This paper addresses constraint solving over continuous domains in the context of decision making, and discusses the trade-off between precision in the definition of the solution space and the computational efforts required. In alternative to local consistency, we propose maintaining global hull-consistency and present experimental results that show that this may be an appropriate alternative to other higher order consistencies. We tested various global hull enforcing algorithms and the best results were obtained with the integration of a local search procedure within interval constraint propagation.

1   Introduction

Model-based decision support relies on an explicit representation of a system in some domain of interest. Given the inevitable simplification that a model introduces, rather than obtaining the solution in the (simplified) model that optimises a certain goal, a decisor is often more interested in finding a compact representation of the solutions that satisfy some constraints. From the extra knowledge that the decisor might have, the set of acceptable solutions may be subsequently focussed into some region of interest defined by means of additional constraints.

An interesting implementation of these ideas [16] addresses this type of problems in the domain of engineering, where models are specified as a set of constraints on continuous variables (ranging over the reals). Such constraints are usually non linear, making their processing quite difficult, as even small errors may easily be expanded throughout the computations. The approach adopted uses octrees, and exploits them in order to determine the regions of the 3D space that satisfy the intended constraints. Although the solution proposed is rather ingenuous, its application to problems with non-convex solution spaces runs into difficulties, namely the large number of convex solutions that might be computed (see [8] for details)

A simpler approach relies on finding upper and lower bounds of every variable that contains the solution space. Although less precise than the former, subsequent interaction with the user may shorten these bounds to specific regions of interest. This approach relies on reasoning about variables that lie on certain intervals, namely the ability to shorten their interval domain by pruning the (outward) regions where it is

C. Bliek et al. (Eds.): COCOS 2002, LNCS 2861, pp. 178–193, 2003.

possible to prove that no solutions may exist. This is the aim of interval constraints, where variables range over intervals and are subject to constraint propagation to reduce their domains.

Constraint propagation of intervals maintains some form of local consistency of the constraint set, typically box-consistency [1, 24] or hull-consistency [18, 2]. Neither is complete and the pruning of domains that is obtained is often quite poor. To improve it, maintenance of higher order local consistencies (3B-consistency [18], Bound-consistency [22]) were proposed. In alternative, we introduced global hull-consistency [6, 9] and applied it to constraint problems with parametric ordinary differential equations, but its initial implementation suffered from various efficiency difficulties.

In this paper, we present improved algorithms to maintain global hull-consistency, and compare them with algorithms that enforce higher order consistencies. Global hull-consistency is in general a suitable alternative, the best results being obtained with the integration of constraint propagation with local search.

The paper is organised as follows. Section 2 overviews the main concepts involved in constraint solving over continuous domains and interval constraints. Section 3 discusses local consistency and higher order alternatives, including global hull-consistency. Section 4 describes several algorithms to enforce it, and section 5 discusses a local search procedure that can be integrated in them. Section 6 presents experimental results obtained with our global hull algorithms and compares them with the competing higher order alternatives. Finally, the main conclusions are summarised and future research directions discussed.

2   Continuous Constraint Solving Problems

A constraint satisfaction problem (CSP) [19] is defined by a set of variables ranging over some domain and a set of constraints. A constraint specifies which values from the domains of its variables are compatible. A solution is an assignment of values to all variables, satisfying all constraints. In continuous CSPs (CCSPs) [16] all domains are continuous real intervals and constraints are equalities or inequalities.

In practice, computer systems may only represent finite subsets of real numbers, called the floating point numbers. We will denote by F-numbers a subset of floating point numbers forming a grid with a specified precision for the representation of the real numbers. An F-interval is a closed real interval whose bounds are F-numbers. It is called *canonical* if the bounds are two consecutive F-numbers. The F-interval approximation of a real interval is the smallest F-interval containing all its elements.

A Real box is the Cartesian product of real intervals. An F-box is the Cartesian product of F-intervals, which is *canonical* if all the F-intervals are canonical.

A *canonical solution* of a CCSP is any canonical F-box that cannot be proved inconsistent (wrt to the CCSP) either because it contains solutions or due to approximation errors in the evaluation of the constraint set.

3   Partial Consistencies: From Local to Global

The Interval Constraints framework, firstly introduced by Cleary [4], combines Artificial Intelligence techniques with Interval Analysis methods for solving CCSPs. The pruning of variable domains is based on constraint propagation techniques initially developed in Artificial Intelligence for finite domains, which use the partial information expressed by a constraint to eliminate incompatible values from the domain of its variables. The reduction of the domain of a variable is propagated to all constraints with that variable in their scopes, which may further reduce the domains of the other constrained variables. Propagation terminates when a fixed point is attained, that is, the variable domains cannot be further reduced by any constraint.

To reduce the domains of variables appearing in a constraint, an appropriate local filtering algorithm (often using Interval Analysis techniques, e.g. the interval Newton method [20, 15]) must enforce a local property, local consistency, relating their domains according to the constraint.

3.1   Local Consistency

The two most common local consistencies used in CCSPs, hull-consistency [2] (or 2B-consistency [18]) and box-consistency [1,24], approximate to arc-consistency[19], widely used in finite domains. Arc-consistency eliminates a value from a variable domain if no compatible value exists in the domain of another variable sharing the same constraint. In continuous domains such enumeration is no longer possible and both hull and box-consistency assume that the domains of the variables are convex (and represented by F-intervals), so they simply aim at tightening their outer bounds.

Hull-consistency guarantees arc-consistency only at the bounds of the variable domains. When a variable is instantiated to one of its bounds, there must be a consistent instantiation of the other constraint variables. Existing algorithms decompose the original constraints into a set of primitive constraints (with addition of extra variables), and the property can be enforced by interval arithmetic operations.

The major drawback of this decomposition is the worsening of the locality problem. Existence of intervals satisfying a local property on each constraint does not imply the existence of value combinations satisfying simultaneously all of them. When a complex constraint is subdivided into primitive constraints this will only worsen this problem due to the addition of new variables and the consequent loss of the dependency between values of related variables.

Box-consistency guarantees the consistency of each bound of the domain of each variable with the F-intervals of the others. After substituting all but one variable by their interval domains, the remaining variable is narrowed by numerical methods, in particular a combination of interval Newton iterates and bisection [1, 24]. The main advantage of this approach is that it does not require the decomposition into primitive constraints, thus avoiding the amplification of the locality problem.

Nevertheless, the domain pruning obtained by box-consistency may still be insufficient. The reason is that if there are n uncertain variables, it is necessary to handle functions with $n-1$ interval values, whose uncertainty may produce a wide and inaccurate range of possible values for these functions.

Generalizing local consistency from a constraint to the set of constraints of a CCSP, we can say that an F-box is Local-consistent wrt a CCSP iff it is locally consistent wrt all the constraints of the CCSP. Application of the (local consistency enforcing) propagation algorithm to an F-box, results in the largest sub-box Local-consistent wrt each constraint of the CCSP.

3.2   Higher Order Consistencies

Better pruning of the variable domains may be achieved if, complementary to a local property, some (global) properties are also enforced on the overall constraint set. 3B-consistency [18] and Bound-consistency [22] are higher order generalizations of hull and box-consistency respectively that enforce the following property on the overall constraint set: if one variable is instantiated to one of its canonical bound, there is a Local-consistent (hull or box-consistent) sub-box wrt the CCSP. Local consistency can be seen as a special case of kB-Consistency (with k=2) formally defined as:

Definition (kB-Consistency). Let P be a CCSP with n variables, B an F-box representing the domains of the variables and k an integer between 3 and $n+1$.

B is 2B-Consistent wrt P iff B is Local-consistent wrt P.

B is kB-Consistent wrt P iff, when the domain of any variable is reduced to one of its (canonical) bounds, there is a sub-box (k-1)B-Consistent wrt P.   ❑

3.3   Global Hull-Consistency

The pruning of the search space achieved by local consistency is often poor, but the computational cost of enforcing stronger consistencies (e.g. by interleaving constraint propagation and search techniques [18, 3]) may limit its practical applicability.

Global Hull-consistency is a strong consistency criterion that we propose as a convenient trade-of. The key idea is to generalise local Hull-consistency to a higher level, by considering the set of all constraints as a single global constraint. Hence, if a variable is instantiated to one of its bounds, there must be a consistent instantiation of the other variables, and this complete instantiation is a solution of the CCSP.

If real values could be represented with infinite precision, Global Hull-consistency is similar to e-consistency (e- for external) [5]. However, due to limitations of the representation of real values, e-consistency cannot be enforced in general.

Definition (Global Hull-Consistency). Let P be a CCSP and B an F-box representing the domains of the variables appearing in P. B is Global Hull-consistent wrt P iff for each bound of each variable domain in B there is a canonical F-box instantiation within B which is a canonical solution of P. ❑

Any strategy to enforce Global Hull-consistency must be able to localise the canonical solutions within a box of domains that are extreme with respect to each bound of each variable domain. Global Hull-consistency is the strongest criterion for narrowing a box of domains into a smaller F-box that looses no possible solution.

4   Enforcing Global Hull-Consistency

The existing constraint systems developed for continuous domains are able to enforce some kind of partial consistency and use it for isolating solutions of a CCSP. Thus, the ultimate goal of these systems is to find individual solutions and not to enclose the complete solution space within a single box.

In CCSPs with a small number of solutions, finding them all could be used for enforcing Global Hull-consistency (the resulting box would enclose the complete set of solutions). In under-constrained CCSPs, the huge number of solutions makes this strategy inadequate and specialised algorithms are needed for enforcing Global Hull-consistency within reasonable computational costs.

4.1   The Higher Order Consistency Approach

For CCSPs with a single variable, enforcing Local-consistency (2B-Consistency) is sufficient to guarantee that each bound is a canonical solution of the CCSP and the resulting box is Global Hull-consistent. In fact, if the variable is instantiated to its leftmost/rightmost canonical subinterval, each constraint must be satisfied. Since there is only one variable in the CCSP, its instantiation is a complete instantiation satisfying all constraints and, consequently, is a canonical solution of the CCSP.

The above property may be generalised for any number of variables occurring in a CCSP: if a CCSP contains n variables, the resulting box obtained by enforcing $(n+1)$B-Consistency is Global Hull-consistent.

4.2   Ordered Search Approaches

Most interval constraint systems provide a search mechanism (alternating pruning and branching steps) that implements a backtracking search for canonical solutions. The pruning is usually achieved by enforcing some Local-consistency (or a stronger kB-consistency), and branching chooses some variable (according to a split strategy) and considers separately the boxes obtained by dividing its domain at the mid point.

To enforce Global Hull-consistency, the ultimate goal is not merely to find canonical solutions but those most extreme with respect to each variable bound. Hence, rather than using the natural backtracking sequence, enforcing algorithms might anticipate the exploration of preferable space regions, thus compensating the extra computational cost of such strategy.

The following subsections describe three different enforcing algorithms which use a mechanism for searching new canonical solutions that keeps the search space ordered with respect to a variable bound. All algorithms maintain an inner box, the smallest box enclosing all the currently found canonical solutions, and an outer box enclosing the whole space of possible solutions. Hence, the algorithms constrain the search to the space between these two boxes. When they merge into a single box, all extreme canonical solutions have been found and the box is returned as the largest Global Hull-consistent F-box.

The inner box is initialised to the solution found by the first call of the generic search mechanism (if there are no canonical solutions, the empty set is returned). The outer box is initialised with the initial variable domains and is updated accordingly to the *best F*-box of the ordered search space.

4.2.1   The OS_1 Algorithm

The OS_1 algorithm separately searches extreme canonical solutions wrt each variable bound. To obtain an extreme canonical bound, canonical solutions are recursively searched, maintaining the search space ordered relatively to that bound.

To avoid searching regions within the inner box, a new constraint is added whenever a new canonical solution is found, imposing the value of the variable whose leftmost/rightmost bound is being searched to be smaller/larger than its leftmost/rightmost value in the updated inner box.

Before every search for a new extreme canonical bound, the search space must be reinitialised in order to disregard all the previous additional constraints that were considered for restraining the search space in other search contexts.

4.2.2   The OS_2 Algorithm

The key idea of the algorithm OS_2 (suggested by Frédéric Benhamou in a personal communication) is to control the branching strategy to direct the search towards the extreme canonical solutions. Instead of constraining the search space whenever a new canonical solution is found, the branching strategy guarantees that the first solution found is already an extreme canonical solution with respect to some variable bound.

Again the search is separately executed for each bound of each variable domain. The search for an extreme canonical bound of some variable x_i, uses a branching strategy that always splits the domain of x_i until it becomes canonical. Since the search space is ordered by the bound of x_i, this strategy assures that the domain of x_i in the *best* box is *better* than that in the *second best* box. When the *best* box is narrowed to a canonical solution, it must thus be an extreme canonical bound of x_i.

4.2.3   The OS_3 Algorithm

In the OS_3 algorithm extreme canonical solutions for each variable domain bound are searched simultaneously within a round robin scheme. The basic idea is to use each call of the generic search mechanism to eliminate at least half the gap between the inner and the outer boxes (wrt a variable bound).

When the search is directed towards a bound of a variable x_i, only the relevant search space, between the inner and the outer boxes (relatively to the bound), is considered and represented by two F-boxes obtained by splitting the x_i domain at its mid point.

This strategy forces the generic search mechanism to explore firstly the best half of the relevant region. If a canonical solution is found then the inner box is enlarged to include it and the other half will be considered no further. Otherwise, the best half is eliminated from the search space. In either case, the gap between the inner and the outer boxes (wrt a variable bound) is at least halved.

4.3   The Tree Structured Algorithm

In all previous algorithms the search to a different variable bound does not take full advantage from the search for previous bounds. The main reason is that the data structures used for representing the search space are optimised during the search for a particular bound and cannot be easily reused in the search for different bounds. Alternatively, a different data structure for the representation of the search space may be used and simultaneously optimised for each variable bound, enabling an efficient search on any of these dimensions. This is the case of the *TSA* algorithm, that uses a tree structured representation of the complete search space and a set of ordered lists, one for each variable bound, to keep track of the relevant *F*-boxes and the respective actions that must be executed on them.

4.3.1 The Data Structures

Besides the inner box, the basic data structures maintained by the algorithm are a binary tree and a vector of ordered lists. Each node of the binary tree is an *F*-box representing a sub-region of the search space that may contain solutions of the CCSP. By definition, a parent box is the smallest *F*-box enclosing its two children. The union of all leaves of the binary tree determines thus the complete search space and the root of the tree defines the smallest *F*-box enclosing it (the outer box).

The vector of ordered lists has $2n$ elements (where n is the number of variables of the CCSP) each being an ordered list associated with each bound of each variable. Each element of a list is a pair (*F*-box, *Action*) where *Action* is a label representing the next action (PRUNE, SEARCH or SPLIT) to perform on the *F*-box. The list associated with the left (right) bound of variable x_i maintains the leaves of the binary tree in ascending (descending) order of their left (right) bounds for the x_i domain.

During the execution of the algorithm, the data structures may be modified either by finding a new canonical solution or by changing the tree as a consequence of deleting, narrowing or splitting one of its leaves.

Relevance regarding a variable bound is a key concept in the enforcing algorithm. Only regions of the search space outside the inner box are relevant. Hence, before exploring any *F*-box, it is necessary to check whether it includes some relevant region for some variable bound, and eventually to extract it from the *F*-box.

Whenever a new canonical solution is found, the inner box must be enlarged to enclose it, and the ordered lists associated with each variable bound must be updated to remove irrelevant elements. Moreover, if the relevant search region of some element is narrowed, the next action to be executed over it must be a prune (propagating the domain reduction of the search region).

When a leaf is deleted from the tree, any associated element in the ordered lists must also be removed. When a leaf is narrowed, any associated element may be reordered and eventually eliminated if it becomes irrelevant. When a leaf is split into two new leaves, then its associated elements must be removed and two new elements considered for insertion in the lists (with action label PRUNE).

4.3.2 The Actions

The *TSA* algorithm alternates prune, search and split actions performed over specific sub-regions (*F*-boxes) of the search space. The pruning of an *F*-box is achieved by enforcing a kB-Consistency criterion. The search action aims at finding a canonical solution within an *F*-box (previously pruned) and may be implemented as a simple check of an initial guess or as a more complete local search procedure (see next section). The split of an F-box (previously searched) is done by splitting one of its variables domains at the mid point.

Any action is performed within a context (the leftmost/rightmost bound of a variable x_i) and executed over some leaf B_t of the binary tree (in particular over the sub-box B relevant in the context). Its consequences must be propagated throughout the data structures maintained by the algorithm.

In the prune action, the kB-Consistency requirement is enforced over B. If the result is the empty set then B is discarded from the binary tree: if B is the whole leaf B_t, then it must be deleted; if it is only a part of B_t, this leaf must be by removing B from B_t. When prune does not narrow B, the next action to be executed over leaf B_t in the same context must be a search. Otherwise (sub-box B is narrowed to some non-empty B'), either B is the whole leaf B_t, which must be narrowed to B', or B is a fraction of the leaf B_t, which must be branched into B' and the F-box obtained by removing B from B_t. In either case, the next action for the new leaf B' is search.

If a canonical solution is found during the search action, the inner box must be updated and its consequences processed as described in the previous subsection. Otherwise, the next action to perform over the leaf B_t in the current context is split.

In the split action, the variable of sub-box B with the largest domain is chosen for split at its mid point. The splitting of the leaf B_t, at the same variable and point, is processed as described in the previous subsection.

5   Local Search

An important characteristic of the *TSA* algorithm is to explore only relevant sub-regions of the search space outside the inner box. Finding a new canonical solution thus reduces such space by enlarging the inner box. Local search techniques for finding canonical solutions may thus significantly enhance the algorithm efficiency.

The key idea of local search techniques is to navigate through points of the search space, and inspecting some local properties of the current point to choose a nearby point to jump to. We developed the following local search algorithm for finding canonical solutions of a CCSP within a search box *FB*:

(i)   Initially a starting point is chosen to be the current point. If the goal is to find the lower (upper) bound of variable x_i, the point is the mid point of *FB* except that the ith domain is the smallest (largest) x_i value within the search box;

(ii)   If the current point is within a canonical solution then the algorithm stops;

(iii)   Otherwise, a multidimensional vector is obtained based on the Newton's method for multidimensional root finding (see next subsection);

(iv)   A minimization process (described in subsection 5.2) obtains a new point inside the search box and within the line segment defined by the current point and the point obtained by applying the multidimensional vector to the current point.

(v)   If the distance between the new point and the current point does not exceed a convergence threshold then the algorithm stops without finding any solution;

(vi)   Otherwise the current point is updated to the new point and the algorithm resumes in step (ii);

The algorithm is guaranteed to stop since step (iv) ensures the minimization of a function for which any canonical solution is a zero and the convergence to a local minimum is detected in step (v).

5.1   Obtaining a Multidimensional Vector – The Newton-Raphson Method

The ultimate goal of obtaining a multidimensional vector is to find a solution of a CCSP by applying it to the current point. The problem of finding a solution of a CCSP is thus converted into the problem of finding a root of a multidimensional vector function F, which can be tackled by an appropriate numerical method, as the Newton-Raphson Method [21, 23].

If a vector function F is defined such that each element F_i is zero iff some constraint of the CCSP is satisfied then, any zero of F satisfies all the constraints.

In case of equality constraints, the associated element on the vector function may be defined by the real expression of the left hand side of the constraint. For example, if the CCSP has two equality constraints $c_1 \equiv x_1^2 + x_2^2 - 1 = 0$ and $c_2 \equiv x_1 \times (x_2 - x_1) = 0$ then $F_1 \equiv x_1^2 + x_2^2 - 1$ and $F_2 \equiv x_1 \times (x_2 - x_1)$ define a vector function F whose zeros are solutions of the CCSP. If there are inequality constraints then they will only be included in the vector function if they are not satisfied in the current point.

Once obtained F, a multidimensional vector δx (the Newton vector) corresponding to one step of the Newton-Raphson iterative method for multidimensional root finding, is computed. The Newton-Raphson method is known to rapidly converge to a root from a sufficiently good initial guess [21]. Hence, applying δx to current point x reaches a root of the multidimensional function $F(x + \delta x) = 0$. Expanding the function in Taylor series and neglecting the higher order terms: $F(x + \delta x) = F(x) + J \cdot \delta x$ (J is the Jacobian matrix). Finally if $x + \delta x$ is a root, then $J \cdot \delta x = -F(x)$.

Solving this equation in order to δx we thus obtain a estimation of the corrections to the current point x that move each function $F_i(x)$ closer to zero simultaneously. To solve the equation, we use a numerical technique, Singular Value Decomposition (SVD), which can be used to solve any set of linear equations (the numerical aspects of the SVD technique are discussed in [13] and practical implementation in [10]).

If the equation has one or more solutions, the SVD technique returns the vector corresponding to the smallest correction to zero F. Among the solutions found for the system, that involving smaller corrections is chosen since the quality of the approximation given by the equation is probably better. If the equation has no solutions, the least-squares solution of minimal norm computed by the SVD is the smallest vector from those giving the best possible correction to a zero of F.

5.2   Obtaining a New Point

One problem of the Newton's iterative method is that it may fail convergence if the initial guess is not good enough. To correct this we followed a globally convergent strategy that guarantees some progress toward the minimisation of $|F|$ at each iteration. Such strategies are often combined with the Newton-Raphson method originating the modified Newton method. While such methods can still fail (by converging into a local minimum of $|F|$) they often succeed where the Newton-Raphson method alone fails convergence.

The fact that the Newton vector δx defines a descent direction for $|F|^2$ guarantees that it is always possible to obtain along that direction a point closer to a zero of F. This new point must lie in the segment defined by $x+\lambda \cdot \delta x$ with $\lambda \in [0..1]$. The strategy to obtain the new point consists of trying different λ values, starting with the largest possible value without exceeding the search box limits, and backtracking to smaller values until a suitable point is reached. If x is close enough to the solution then the Newton step has quadratic convergence. Otherwise a smaller step in the direction of a solution (or a mere local minimum) is taken, guaranteeing convergence.

In alternative to the Newton approach, other minimization methods (as the Conjugate Gradient Method [11, 12]) could have been directly applied to the scalar function $|F|^2$. However, the early collapsing of the various dimensions of F into a single one implies the lack of information about each individual constraint and makes these strategies more vulnerable to local minima.

5.3   Integration of Local Search with Global Hull-Consistency Algorithms

The local search algorithm was originally conceived for providing the *TSA* algorithm with a specialised method for the localisation of new canonical solutions within a search box. Hence, local search is naturally integrated as a step of the search action. The integration of local search with the other enforcing algorithms presented in the previous section requires minor changes in their generic search mechanisms.

After enforcing kB-consistency for pruning the variable domains of the *best F*-box, the generic search mechanism should perform a local search for finding a new canonical solution. If a new solution is found and the algorithm is not the OS_2 algorithm then the backtracking search terminates returning the new canonical solution. In the case of OS_2, the backtrack search only terminates if the new solution is an extreme bound; otherwise two new F-boxes must be included in the ordered search space: the new canonical solution and the sub-box that remains relevant after considering this new solution.

6   Experimental Results

In this section we present results obtained by imposing Global Hull consistency (GH) in a number of problems and compare them with those obtained with (various levels of) higher order consistency. The GH algorithm used for these comparisons is TSA with Local Search (precision $\varepsilon = 10^{-3}$), and in the end of the section we compare its efficiency with the other algorithms presented in the previous sections. We implemented all GH and higher order consistency (kB [18, 3]) algorithms, using the procedures for achieving 2B-consistency (box-consistency) available in the OpAC library (a C++ interval constraint language [14]), and executing them on a Pentium III computer at 500 MHz with 128 Mbytes memory.

6.1   A Simple Example

To illustrate the pitfalls often arisen with kB-consistency we may consider a small problem consisting of two constraints:

$$x^2 + y^2 <= 1 \quad \text{and} \quad x^2 + y^2 >= 2$$

Of course, the two constraints are unsatisfiable. However, 2B-consistency does not detect such inconsistency, merely pruning the initial unbounded domains of the variables to the interval [-1.001 .. 1.001]. With these domains (obtained by applying 2B to the first constraint) the second constraint does not prune the domain of any of its variables when the other is fixed to any of its bounds. Of course, x = ±1 is only compatible with y = 0 in the first constraint whereas it requires y = ± 1 in the second, but the local nature of 2B-consistency, does not detect this situation.

Since there are only 2 variables involved, 3B-consistency is equivalent to Global Hull consistency and detects the inconsistency. With an increased number of variables the insufficiency of 2B also arises in higher order consistencies. For example, with

$$x^2 + y^2 + z^2 <= 2 \quad \text{and} \quad x^2 + y^2 + z^2 >= 3$$

rather than detecting inconsistency, as GH does, 3B consistency prunes the variables to [-1.001 .. 1.001] whereas 2B consistency performs even worse pruning the domains to [-1.416 .. 1.415]. In general, the difference obtained with kB-consistency and Global Hull is not so significant, but still different bounds are obtained. For example with the slightly modified problem

$$x^2 + y^2 + z^2 <= 2 \quad \text{and} \quad (x-0.5)^2 + y^2 + z^2 >= 2.25$$

the results obtained are shown in Table 1. Both 2B and 3B, although faster than GH, report quite inaccurate upper bounds for variable x.

6.2   The Census Problem

The Census problem models the variation with time of a population with limited growth by means of a parametric differential equation (logistic). The equation has an

analytical solution of the form: $$x(t) = \frac{kx_0 e^{rt}}{x_0 (e^{rt} - 1) + k}$$

Given a set of observations at various time points, the goal of the problem is to adjust the parameters x_0, r and k of the equation to the observations. We used the US Census over the years 1790 (normalised to 0) to 1910 with a 10 year period. This is done by imposing the above constraint at the time points of Table 2 (with r multiplied by 0.001 to re-scale it into the interval [1..100] and an acceptable error of ±1).

Table 1. Pruning domains in a trivial problem

	2B	3B	GH
x	-1.416 .. 1.415	-1.415 .. 1.002	-1.415 .. 0.001
y	-1.416 .. 1.415	-1.415 .. 1.415	-1.415 .. 1.415
z	-1.416 .. 1.415	-1.415 .. 1.415	-1.415 .. 1.415
t (ms)	10	50	1860

Table 2. US Population over the years 1790 (0) to 1910 (120)

t	0	10	20	30	40	50	60	70	80	90	100	110	120
x	3.929	5.308	7.239	9.638	12.86	17.06	23.19	31.43	39.81	50.15	62.94	75.99	91.97

Table 3. Comparing 2B, 3B and GH in the Census problem

	2B	3B	GH
xo	2.929 .. 4.930	2.929 .. 4.862	3.445 .. 4.547
k	1.1 .. 1000	102.045 .. 306.098	166.125 .. 260.401
r	1.1 .. 100	27.474 .. 39.104	28.683 .. 33.714
t(ms)	10	56 990	756 460

Table 4. Comparing anytime GH and 4B in the Census problem

	GH s	GH s1	GH s2	4B
xo	3.445 .. 4.547	3.020 .. 4.768	3.142 .. 4.566	3.408 .. 4.548
k	166.125 .. 260.401	130.684 .. 283.341	148.242 .. 261.736	164.412 .. 260.632
r	28.683 .. 33.714	27.838 .. 36.625	28.630 .. 35.229	28.677 .. 33.896
t(ms)	756 460	15 070	55 220	1 993 770

Table 5. Square distances between pairs of atoms of the protein

a1-a2	a1-a3	a1-a5	a1-a6	a2-a3	a2-a4	a2-a5	a3-a4	a3-a5	a3-a6	a4-a5	a4-a6
2	4	4	4	2	4	2	2	4	4	2	2

In Table 3 we show the results of enforcing 2B, 3B, and GH on this problem. The table shows the poor pruning results achieved by 2B consistency alone, and the much better pruning achieved by GH wrt 3B consistency.

Although this improvement apparent requires a much longer execution time, it is important to notice that OS_3 and *TSA* algorithms for achieving GH are anytime algorithms, and good results may be obtained much earlier. Table 4 shows that the pruning achieved by GH, at approximately 30% of the execution time spent with 3B, is already significantly better than it. For similar execution times, the pruning is almost as good as the final one. Also notice that, since there are only 3 variables, imposing 4B provides the same results than GH. However, as shown in the table, the algorithm is much slower (it was interrupted at the time reported in the table).

6.3   Protein Structure

The final problem is a simplification of that of finding the structure of a protein from distance constraints, e.g. obtained from Nuclear Magnetic Resonance data (see [17]). The simplified problem uses Euclidean distance constraints similar to those presented in section 6.1 above, where variables x_i, y_i and z_i represent the centres of atom a_i. In this problem, we place 6 atoms, whose centres must all be 1Å apart. For some atom pairs, the square of the distances are provided and shown in table 5.

To solve the problem we placed 3 atoms (a1-a3) arbitrarily in the XY plane. The distances allow that the other 3 atoms are either above the plane (positive Z) or the corresponding quiral solution below the plane (negative Zs). In table 6 we show the pruning achieved on the values of atoms a4 to a6 by 2B, 3B and GH (as before, results with 4B are similar to those with GH but take much longer to obtain).

Given this uncertainty in the Z value of the atoms, neither 2B nor 3B could prune the size of their other dimensions to the amount that GH does. The most important difference between 3B and GH lies on atom a6. Whereas the x6 and y6 are "fixed" (respectively at around −1 and 1) by GH, 3B could not prune the value of these variables beyond the typical −1..1 interval.

6.4   Local Search

In all the above experiments GH was enforced with the TSA algorithm with local search, a precision $\varepsilon = 10^{-3}$ and an underlying 2B-consistency procedure. To check whether the above timings were truly representative of GH enforcement, we decided to test the efficiency of the different algorithms presented in sections 4 and 5. The results presented in Table 7 refer to another instance of the Protein structure (with 8 atoms), more representative of the kind of problems GH is aimed at, in that they have many adjacent solutions, i.e. the final domains of all variables are relatively large.

The first thing to notice is the importance of the precision used. Of course, with less precision, all algorithms are faster, since less canonical F-boxes are considered. More interestingly, as precision increases (smaller ε) local search becomes more important. The reason for this is that with larger canonical F-boxes the underlying 2B algorithm does not prune them and so accepts them as canonical solutions, making the local search for "real" solutions useless. With higher precision, canonical F-boxes are smaller and the 2B algorithm does not detect solutions as easily, since the pruning of most F-boxes does not result into canonical F-boxes. Hence, the advantage of local search in such situations.

Local search also proved to improve the memory requirements, since it often finds solutions near the intended bounds of the variables in the F-boxes under consideration, rather than simply bisecting them.

Table 6. Comparing 2B, 3B and GH in the Protein problem

	2B	3B	GH
x_4	−1.001 .. 1.415	−0.056 .. 0.049	−0.004 .. 0.004
y_4	0.585 .. 3.001	1.942 .. 2.047	1.996 .. 2.004
z_4	−1.415 .. 1.416	−1.415 .. 1.415	−1.415 .. 1.415
x_5	−0.415 .. 2.001	0.998 .. 1.002	0.999 .. 1.001
y_5	−0.001 .. 2.001	0.999 .. 1.001	0.999 .. 1.001
z_5	−1.415 .. 1.416	−1.415 .. 1.415	−1.415 .. 1.415
x_6	−2.001 .. 2.001	−1.110 .. 1.053	−1.008 .. −0.992
y_6	−0.001 .. 2.001	−0.894 .. 1.169	0.999 .. 1.001
z_6	−2.001 .. 2.001	−1.483 .. 1.483	−1.420 .. 1.402
t (ms)	10	7 380	62 540

Table 7. Comparing various GH enforcing algorithms

	k	LS	Time (s)			Max Storage (F-boxes)		
			$\varepsilon=10^{-1}$	$\varepsilon=10^{-2}$	$\varepsilon=10^{-3}$	$\varepsilon=10^{-1}$	$\varepsilon=10^{-2}$	$\varepsilon=10^{-3}$
OS_1	2	n	7.34	600+	600+	407	55711	63465
		y	87.52	78.48	600+	407	414	25580
	3	n	23.57	139.84	600+	27	339	1584
		y	12.80	50.39	191.41	2	15	51
OS_2	2	n	2.99	11.37	600+	30	47	60
		y	14.97	38.46	600+	14	16	24
	3	n	41.01	150.21	359.69	22	37	53
		y	19.14	75.98	234.65	3	8	25
OS_3	2	n	5.19	600+	600+	132	66410	63445
		y	23.29	20.72	600+	51	104	25037
	3	n	25.51	185.99	600+	19	695	1676
		y	13.08	45.41	162.50	2	10	32
TSA	2	n	10.16	60.96	600+	332	6786	76506
		y	44.79	47.59	600+	221	621	16764
	3	n	42.52	184.38	600+	52	207	843
		y	38.63	97.88	275.85	99	185	392

Memory requirements are also much lower when the underlying enforcement procedure is 3B (rather than 2B), as the pruning achieved in any F-box is much more significant. Because of its higher pruning, and despite its higher complexity, procedure 3B provides in general better results than 2B, as precision increases.

Regarding the variety of GH enforcing algorithms, their differences are less evident. Given the discussion above it might be important to impose some thresholds on the execution time of the algorithms, in which case the OS_1 and OS_2 have the disadvantage of not being anytime algorithms. Although OS_3 proved better than TSA in this problem, this behaviour is not observed consistently in other problems, and TSA offers the advantage of keeping a tree-based compact description of the feasible space, which is very convenient for an interactive use as envisaged in [16]. Moreover, and although not visible in the table, the anytime results of TSA are consistently better than those obtained with OS_3.

7   Conclusions

This paper addressed constraint solving over continuous domains, and shows that our proposed approach to enforce global hull consistency may be an appropriate choice, compared with other higher order consistency requirements, namely kB-consistency.

The paper compares experimental results obtained with various global hull enforcing algorithms, and shows the importance of local search to improve them, especially when more precision is required.

Although the results were obtained in a set of relevant examples, we intend to use it in more complex and realistic problems, namely to improve our previous work on continuous constraints with differential equations [7], and to integrate it in the more demanding problem of determining the structure of proteins from NMR data [17].

Acknowledgements

This work was partly supported by project PROTEIN – POSI/SRI/33794/99 granted by the Portuguese Foundation for Science and Technology. Special thanks are due to F. Benhamou, for the valuable discussions.

References

[1]   Benhamou, F., McAllester, D., Van Hentenryck, P.: CLP(Intervals) revisited. *Logic Programming Symposium*, MIT Press (1994) 124-131.
[2]   Benhamou, F., Older, W.J.: Applying Interval Arithmetic to Real, Integer and Boolean Constraints. *Journal of Logic Programming* (1997) 32(1): 1-24.
[3]   Bordeaux, L., Monfroy, E., Benhamou, F.: Improved bounds on the complexity of kB-consistency. Proceedings of 17th IJCAI, (2001) 640-650.
[4]   Cleary, J.G.: Logical Arithmetic. *Future Computing Systems* (1987) 2(2):125-149.
[5]   Collavizza, H., Delobel, F., Rueher, M.: Extending Consistent Domains of Numeric CSP. IJCAI, (1999) 406-413.
[6]   Cruz, J., Barahona, P.: An Interval Constraint Approach to Handle Parametric ODEs for Decision Support. *Principles Practice Constraint Programming*, Springer (1999) 478-479.
[7]   Cruz, J., Barahona, P., Benhamou, F.: Integrating Deep Biomedical Models into Medical DSSs: an Interval Constraint Approach. *AI in Medicine*, Springer (1999) 185-194.
[8]   Cruz J., Barahona, P.: Global Hull Consistency with Local Search for Continuous Constraint Solving. 10th Portuguese Conference on AI, Springer, (2001) 349-362.
[9]   Cruz, J., Barahona, P.: Handling Differential Equations with Constraints for Decision Support. *Frontiers of Combining Systems*, Springer (2000) 105-120.
[10]   Demmel, J.W.: Applied Numerical Linear Algebra. SIAM, Philadelphia, PA, (1997).
[11]   Dennis, J.E., Schnabel, R.B.: Numerical Methods for Unconstrained Optimization and Nonlinear Least Squares. Prentice-Hall, Englewood Cliffs, NJ, (1983).
[12]   Fletcher, R., Reeves, C.: Function Minimization by Conjugate Gradients. Computer Journal, (1964) 7:149-154.

[13] Golub, G.H., Van Loan, C.F.: Matrix Computations, 3th Ed. Johns Hopkins University Press, Baltimore and London, (1996).

[14] Goualard, F.: Langages et environnements en programmation par contraintes d'intervalles. Thèse de doctorat. Université de Nantes, (2000).

[15] Hansen, E.: Global Optimization Using Interval Analysis. Marcel Dekker (1992).

[16] Sam-Haroud, D., Faltings, B.V.: Consistency Techniques for Continuous Constraints. *Constraints* (1996) 1(1,2):85-118.

[17] Krippahl, L., Barahona, P.: Applying Constraint Propagation to Protein Structure Determination, *Principles Practice Constraint Programming*, Springer (1999) 289-302.

[18] Lhomme, O.: Consistency Techniques for Numeric CSPs. *IJCAI*, IEEE Pr. (1993) 232-238.

[19] Montanari U.: Networks of Constraints: Fundamental Properties and Applications to Picture Processing. *Information Science* (1974) 7(2):95-132.

[20] Moore, R.E.: Interval Analysis. Prentice-Hall (1966).

[21] Ortega, J., Rheinboldt, W.: Iterative Solution of Nonlinear Equations in Several Variables. Academic Press (1970).

[22] Puget, J-F., Van Hentenryck, P.: A Constraint Satisfaction Approach to a Circuit Design Problem. *Journal of Global Optimization*, MIT Press (1997).

[23] Stoer, J., Burlisch R.: Introduction to Numerical Analysis. Springer Verlag (1980).

[24] Van Hentenryck, P., McAllester, D., Kapur, D.: Solving Polynomial Systems Using a Branch and Prune Approach. *SIAM Journal of Numerical Analysis* (1997) 34(2).

Numerical Constraint Satisfaction Problems with Non-isolated Solutions

Xuan-Ha Vu, Djamila Sam-Haroud, and Marius-Calin Silaghi

Artificial Intelligence Laboratory
Swiss Federal Institute of Technology Lausanne (EPFL)
CH-1015, Lausanne, Switzerland
{xuan-ha.vu,jamila.sam,marius.silaghi}@epfl.ch
http://liawww.epfl.ch

Abstract. In recent years, interval constraint-based solvers have shown their ability to efficiently solve complex instances of non-linear numerical CSPs. However, most of the working systems are designed to deliver *point-wise* solutions with an arbitrary accuracy. This works generally well for systems with isolated solutions but less well when there is a *continuum of feasible points* (e.g. under-constrained problems, problems with inequalities). In many practical applications, such large sets of solutions express equally relevant alternatives which need to be identified as completely as possible. In this paper, we address the issue of constructing *concise* inner and outer approximations of the complete solution set for non-linear CSPs. We propose a technique which combines the *extreme vertex representation* of orthogonal polyhedra [1, 2, 3], as defined in computational geometry, with adapted *splitting strategies* [4] to construct the approximations as unions of interval boxes. This allows for compacting the explicit representation of the complete solution set and improves efficiency.

1 Introduction

Many practical problems require solving constraint satisfaction problems (CSPs) with numerical constraints. A numerical CSP (NCSP), $(\mathcal{V}, \mathcal{C}, \mathcal{D})$, is stated as a set of variables $\mathcal{V}$ taking their values in domains $\mathcal{D}$ over the reals and subject to a finitely many set of constraints $\mathcal{C}$. In practice, the constraints can be equalities or inequalities of arbitrary type and arity, usually expressed using arithmetic expressions. In this paper we address the case of NCSPs with *non-isolated solutions*. Such a case is often encountered in real-world engineering applications where under-constrained problems, problems with inequalities or with universal quantifiers are ubiquitous. In practice, a set of non-isolated solutions often expresses a spectrum of equally relevant choices, as the possible moving areas of a mobile robot, the collision regions between objects in mechanical assembly, or different alternatives of shapes for the components of a kinematic chain. These alternatives need to be identified as precisely and completely as possible.

Interval constraint-based solvers (e.g. Numerica [5], ILOG Solver [6]) take as input an NCSP and generate a set of boxes which *conservatively* enclose each

C. Bliek et al. (Eds.): COCOS 2002, LNCS 2861, pp. 194–210, 2003.

solution. They have proven particularly efficient in solving challenging instances of NCSPs with non-linear constraints. However, when applied to problems with non-isolated solutions they provide enclosures that are either prohibitively verbose or poorly informative (see Section 2).

In contrast, a number of set-based approaches have been developed, notably in the areas of robust control, automation and robotics, which provide promising alternatives to the point-wise techniques. They consist in covering the spectrum of non-isolated solutions using a reduced number of subsets of $\mathbb{R}^n$. Usually, these subsets are chosen with known and simple properties (e.g. interval boxes, polytopes, ellipsoids). In recent years, several authors have proposed set covering algorithms with interval boxes [7, 8, 9, 10]. Most existing box-covering algorithms are however limited by their restrictive applicability conditions or by their high average time and space complexities in the general case. The enhanced set-based technique we propose builds on the following observations. Firstly, the union of boxes produced by the complete interval-based solving of NCSPs can be seen as an *orthogonal polyhedron*[1]. Enhanced representations from computational geometry can be used to reduce the verbosity of such geometrical objects. We propose to use the *Extreme Vertex Representation* (EVR) of orthogonal polyhedra [1, 2, 3] for this purpose. Secondly, when there are non-isolated solutions, dichotomous splitting is not the most adapted branching strategy. It might lead to unnecessarily dividing entirely feasible regions. We propose to use another scheme based on splitting around the negation of feasible regions [4] which is an extension of the *negation test* proposed for universally quantified constraints in [10]. The resulting algorithm applies to general constraint systems. It produces inner and outer approximations of the feasible sets in the form of *unions of interval boxes*. The preliminary experiments show that it improves efficiency as well as the compactness and quality of the output representation.

2 Examples

We start by giving two small introductory examples which illustrate the inadequacy of point-wise approaches to the case of NCSPs with non-isolated solutions. The first example illustrates how the point-wise approach can be sometimes misused when applied to NCSPs with non-isolated solutions. Since point-wise techniques inherently assume the existence of isolated solutions, the interval splitting process they use for branching is sometimes prematurely stopped as soon as a solution is detected within an interval. This leads to poorly informative approximations of the complete solutions sets, as shown by the following example. The first example, called **WP**, is a 2D simplification of the design model for a kinematic pair consisting of a wheel and a pawl. The constraints determine the regions where the pawl can touch the wheel without blocking its motion. $\mathbf{WP} = \{20 < \sqrt{x^2 + y^2} < 50,\ 12y/\sqrt{(x-12)^2 + y^2} < 10,\ x \in [-50, 50],\ y \in [0, 50]\}.$

---

[1] Informally, an orthogonal polyhedron is such that its facets are axis-parallel hyperrectangles.

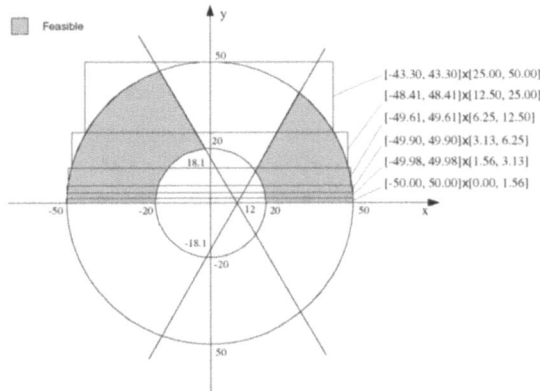

Fig. 1. The solution set of **WP** is approximated by 6 boxes (at precision $= 2$)

Figure 1 shows the output produced by a point-wise solver when the existence of point-wise solutions is abusively assumed.[2]

The second example consists of 4 non-linear inequality constraints involving 3 variables: $\mathbf{P3} = \{x^2 \leq y,\ \ln y + 1 \geq z,\ xz \leq 1,\ x^{3/2} + \ln(1.5z + 1) \leq y + 1,\ x \in [0, 15],\ y \in [1, 200],\ z \in [0, 10]\}$. Using an efficient implementation of classical point-wise techniques,[3] the computation had to be stopped after 10 hours and produced more than 260000 small boxes. The alternative set-based technique we propose could reduce the complete output to 1376 boxes and produced the result in 1.41 seconds (see Table 1). This small example was one of our most successful and hence does not objectively illustrate the power of our technique, however it clearly illustrates how point-wise approaches can be unadapted to the complete solving of certain classes of problems.

3   Background and Definitions

3.1   Interval Arithmetic

The finite nature of computers precludes an exact representation of the *reals*. The real set, $\mathbb{R}$, is in practice approximated by a finite set $\mathbb{F}_\infty = \mathbb{F} \cup \{-\infty, +\infty\}$, where $\mathbb{F}$ is a finitely many set of reals which usually corresponds to the floating-point numbers. For each $x \in \mathbb{F}$, we denote $x^+ = \min\{y \in \mathbb{F}_\infty \mid x < y\}$, $x^- = \max\{y \in \mathbb{F}_\infty \mid y < x\}$. In this paper, we restrict the notion of *interval* to refer to real intervals with bounds in $\mathbb{F}_\infty$, no matter they are open or closed [10]. The set of intervals with bounds in $\mathbb{F}_\infty$, denoted by $\mathbb{I}$, is partially ordered by the relation '$<$' on reals. An *interval box*, or a *box* for short, $\mathbf{B} = I_1 \times \ldots \times I_n$ is

---

[2] It was solved using a combination of *IloGenerateBounds* and *IloSplit* in ILOG Solver 5.2.

[3] The implementation was based on ILOG solver 5.2 (see Section 6).

a Cartesian product of n intervals in $\mathbb{I}$. We denote $\mathbf{B}|_i = I_i$. A *canonical interval* is a non-empty interval, I, such that $\sup(I) \leq \inf(I)^+$. Two boxes, $\mathbf{A}$ and $\mathbf{B}$, are called *disjoint* if $\mathbf{A} \cap \mathbf{B} = \emptyset$. We denotes by $pts(S)$ the set of points represented by S, e.g. $pts(S) = \{x \mid x \in \mathbf{B} \in S\}$ if S is a set of boxes. Two sets of disjoint boxes, $\mathcal{U}_1$ and $\mathcal{U}_2$, are equivalent, denoted by $\mathcal{U}_1 \equiv \mathcal{U}_2$, if $pts(\mathcal{U}_1) = pts(\mathcal{U}_2)$.

3.2   Relations and Approximations

Let $c(x_1, \ldots, x_n)$ be a real constraint with arity n. The *relation* defined by c, denoted by ρ_c, is the set of tuples in $\mathbb{R}^n$ satisfying c. Let $vars(\rho_c) = vars(c) = \{x_1, \ldots, x_n\}$. The relation defined by the negation, $\neg c$, of c is given by $\mathbb{R}^n \setminus \rho_c$. In this paper, each box $\mathbf{B} \in \mathbb{I}^n$ and relation $\rho \subseteq \mathbb{R}^n$ is associated with n real variables, the projection of $\mathbf{B}$ on a subset, X, of its variables is denoted by $\mathbf{B}|_X$. We denote by $\mathcal{R}_n$ the set of relations defined on subsets of the n variables, ρ_P the relation defined by an NCSP, P, and ρ_C the *global relation* defined by the conjunction of constraints in a constraint set C.

A relation can be approximated by a computer-representable superset or subset. The former is a *complete* approximation but may contain points that are not solutions. Conversely, the latter is a *sound* approximation but may lose certain solutions. In particular, a relation can be approximated conservatively by the smallest (w.r.t. set inclusion) union of boxes (called the *best outer approximation*), or more coarsely by the smallest box (called the *interval hull*), containing it. Beside that, the relation can also be approximated by the greatest union of boxes (called the *best inner approximation*) contained in it. The readers are referred to [11] for rigorous definitions.

The computation of these ideal approximations relies on the notion of *contracting operators*. An outer-box contracting operator [10] narrows down the variable domains by discarding values that are locally inconsistent using *Box* consistency. In this paper we use a generic notion defined as follows:

Definition 1 (Outer-Bound Contracting Operator, OC). *An outer-bound contracting operator is a function* $\mathsf{OC} : \mathbb{I}^n \times \mathcal{P}(\mathbb{R}^n) \to \mathbb{I}^n \cup \{\emptyset\}$ *such that* $\forall \mathbf{B} \in \mathbb{I}^n$, $\rho \in \mathcal{P}(\mathbb{R}^n)$ *these properties hold:* [4]

$$(i)\ \mathsf{OC}(\mathbf{B}, \rho) \subseteq \mathbf{B} \qquad (\textit{Contractiveness})$$
$$(ii)\ \mathsf{OC}(\mathbf{B}, \rho) \supseteq \mathbf{B} \cap \rho \qquad (\textit{Completeness})$$

In numerical domains, the outer-bound contracting operators usually enforce either *Box*, *Hull*, *kB* or *Bound* consistency and *monotonicity* [12, 5], generally referred to as bound-consistency in the rest of the paper. For simplicity, given a set of constraints C, we use C instead of ρ_C in the notation of contracting operators.

Proposition 1. *Given a set of constraints, C, and a bounding box, $\mathbf{B}$. The box $\mathbf{B}$ is completely infeasible (w.r.t. C) if there is some* OC *operator that contracts* $(\mathbf{B}, C)$ *to an empty set, i.e.* $\exists \mathsf{OC} : \mathsf{OC}(\mathbf{B}, C) = \emptyset \Rightarrow \mathbf{B}$ *is infeasible (w.r.t. C).*

---

[4] $\mathcal{P}(S)$ denotes the power set of S, i.e., the set $\{A \mid A \subseteq S\}$.

3.3   Union Approximations

In general, the computation of the best inner and outer approximations is intractable. Therefore, in this paper we consider the problem of computing inner and outer approximations of a relation $\rho \subseteq \mathbb{R}^n$ in the form of *unions of disjoint boxes*.

Definition 2 (Outer Union Approximation, Union$^\mathcal{O}$). Union$^\mathcal{O}(\rho)$ *is a set of disjoint boxes* $\mathcal{U} \in \mathcal{P}(\mathbb{I}^n)$ *such that* $pts(\mathcal{U}) \supseteq \rho$.

Definition 3 (Inner Union Approximation, Union$^\mathcal{I}$). Union$^\mathcal{I}(\rho)$ *is a set of disjoint boxes* $\mathcal{U} \in \mathcal{P}(\mathbb{I}^n)$ *such that* $pts(\mathcal{U}) \subseteq \rho$.

Definition 4 (Undiscernible   Union   Approximation,   Union$^\mathcal{U}$). Union$^\mathcal{U}(\rho)$ *corresponding to* Union$^\mathcal{O}(\rho)$ *and* Union$^\mathcal{I}(\rho)$ *is a set of disjoint boxes* $\mathcal{U} \in \mathcal{P}(\mathbb{I}^n)$ *such that* $pts(\mathcal{U}) = pts(\mathbf{Union}^\mathcal{O}(\rho)) \setminus pts(\mathbf{Union}^\mathcal{I}(\rho))$.[5]

Several authors have recently addressed the issue of computing **Union$^\mathcal{O}$** approximations. In [7], a recursive dichotomous split is performed on the variable domains. Each box obtained by splitting is tested for inclusion using interval arithmetic tools. The boxes obtained are hierarchically structured as 2^k-*trees*. The authors have demonstrated the practical usefulness of such techniques in robotics, automation and robust control. In [8], a similar algorithm is presented. However, only binary or ternary subsets of variables are considered when performing the splits. The approach is restricted to classes of problems with convexity properties. The technique proposed in [9] algebraically constructs the unions using Bernstein polynomials which makes it possible to use guaranteed algebraic inclusion tests for boxes. The approach is restricted to polynomial constraints. A technique of extending consistent domains of a particular class of constraints has also been proposed in [13]. Finally, [10] has addressed the issue of computing **Union$^\mathcal{I}$** for universally quantified constraints using *negation tests* (see Section 4.1), and [4] has extended the negation test in combination with enhanced splitting strategies to computing **Union$^\mathcal{O}$** for classic NCSPs. Hereafter, we give abstractions of conventions in computing union approximations where **Union$^\mathcal{U}$** is assumed to exist.

Definition 5 (Feasibility Checker, FC). *A feasibility checker is a function,* $\mathrm{FC} : \mathbb{I}^n \times \mathcal{R}_n \to \{\texttt{feasible}, \texttt{infeasible}, \texttt{unknown}\}$ *such that:*

   (i) $\mathrm{FC}(\mathbf{B}, \rho) = \texttt{feasible} \Rightarrow \mathbf{B}|_{vars(\rho)} \subseteq \rho$

   (ii) $\mathrm{FC}(\mathbf{B}, \rho) = \texttt{infeasible} \Rightarrow \mathbf{B}|_{vars(\rho)} \subseteq \neg\rho$

  (iii) $\mathrm{FC}(\mathbf{B}, \rho) = \texttt{unknown} \wedge \mathbf{B}|_{vars(\rho)} \subset \mathbf{B}'|_{vars(\rho)} \Rightarrow \mathrm{FC}(\mathbf{B}', \rho) = \texttt{unknown}$

Definition 6 (Interval-Based Precision). *Given an NCSP, P, a precision (vector),* ε, *and a feasibility checker,* FC. *A search technique which computes the union approximations is called having the precision* ε *(w.r.t. FC) if there is some set,* $\mathcal{U}$, *of disjoint boxes whose sizes are not greater than* ε *(component-wise) such that:*

$$\mathcal{U} \equiv \mathbf{Union}^\mathcal{U}; \forall \mathbf{B} \in \mathcal{U} : \mathrm{FC}(\mathbf{B}, \rho_P) = \texttt{unknown} \tag{1}$$

---

[5] Informally, **Union$^\mathcal{U}(\rho)$** is a set of undiscernible boxes enclosing the boundary of ρ.

4 EVR and Complementary-Boxing

Interval-based search techniques for NCSPs are essentially bisectional. Variables are instantiated using intervals. When the search reaches an interval that contains no solutions it backtracks, otherwise the interval is recursively split into two halves up to an established resolution. The most successful techniques enhance this process by applying an inclusion test or OC operator to the overall constraint system, after each split. In most known algorithms, the general policy is to perform splitting intervals until canonical intervals are reached or their widths are not greater than a predefined precision, i.e. it simply has a predefined interval-based precision. This policy, referred to as DMBC (dichotomous maintaining bound-consistency) in the rest of the paper, generates verbose outer and inner union approximations. The first reason is that the orthogonal splitting policy introduces artificial convexity deficiencies and generates a significant number of nearly aligned boxes along boundaries of constraints. The second reason is that entirely feasible boxes might be unnecessarily split. The improvements we propose are presented in the two next subsections.

4.1 Better Splitting Decisions Using Complementary-Boxing

We now recall the techniques which construct inner approximations [10] and outer approximations [4] by soon isolating feasible regions under new concepts in order to be integrated with the techniques described in Section 4.2. Given a relation, ρ, and a box, $\mathbf{B}$, the *negation test* performs a kind of OC operator on $(\mathbf{B}, \neg\rho)$. A kind of splitting operator (called ICAb3$_c$) which splits around a box obtained by a negation test and dichotomizes this box was proposed in [10]. The negation approach, which is called ICAb5 and proposed for universally quantified constraints, recursively performs ICAb3$_c$ on the first active constraint until a predefined interval-based precision reached. In [4], a similar splitting operator based on the negation test, which is called $\mathbf{B_q}$ therein, was employed for numeric constraints. However, the approach, which is called UCA6 and proposed for NCSPs, computes negation tests for all constraints and then chooses the best for splitting. In addition, UCA6 memorizes old $\mathbf{B_q}$'s for computing new $\mathbf{B_q}$'s and performed some mixed splitting strategies based on equalities/inequalities. Hereafter, we employ the negation test to define a contracting operator, called *Complementary-Box contracting operator* and a splitting operator, called *Box splitting operator*. Their definitions are given as follows.

Definition 7 (Complementary-Box   Contracting   Operator,   CBC).
A Complementary-Box contracting operator is a function CBC $: \mathbb{I}^n \times \mathcal{P}(\mathbb{R}^n) \to \mathbb{I}^n \cup \{\emptyset\}$ *such that* $\forall \mathbf{B} \in \mathbb{I}^n$, $\rho \in \mathcal{P}(\mathbb{R}^n)$ *these properties hold:*[6]

$$(i)\ \mathrm{CBC}(\mathbf{B}, \rho) \subseteq \mathbf{B} \qquad (Contractiveness)$$
$$(ii)\ \mathbf{B} \setminus \mathrm{CBC}(\mathbf{B}, \rho) \subseteq \rho \ (Complementariness)$$

---

[6] In [11], this operator was named "Back-Boxing Contracting operator". In this paper, we change its name and notation to avoid confusions with the namesake given in [14].

A box resulting from the application of a CBC operator to a bounding box, **B**, and a relation, ρ, is called a *Complementary-Box* with respect to ρ within **B**. *Complementary-Boxing* refers the process of identifying the Complementary-Box. The following properties characterize CBC operators.

Proposition 2. *Given a set of constraints, $\mathcal{C}$, and a bounding box, **B**. The box **B** is completely feasible (w.r.t. $\mathcal{C}$) if there is some CBC operator that contracts $(\mathbf{B}, \mathcal{C})$ to an empty set, i.e. $\exists \mathrm{CBC} : \mathrm{CBC}(\mathbf{B}, \mathcal{C}) = \emptyset \Rightarrow \mathbf{B}$ is feasible (w.r.t. $\mathcal{C}$).*

Proposition 3. *Given an OC operator. The function $f : \mathbb{I}^n \times \mathcal{P}(\mathbb{R}^n) \to \mathbb{I}^n \cup \{\emptyset\}$ defined by $f(\mathbf{B}, \rho) = \mathrm{OC}(\mathbf{B}, \neg\rho)$ is a CBC operator.*

Proof. By definition $\mathbf{B}_f = f(\mathbf{B}, \rho) = \mathrm{OC}(\mathbf{B}, \neg\rho)$. The contractiveness of OC operators implies $\mathbf{B}_f \subseteq \mathbf{B}$, i.e. the contractiveness of CBC operators. In addition to the completeness of OC operators we have $\mathbf{B} \cap \neg\rho \subseteq \mathbf{B}_f \subseteq \mathbf{B} \Rightarrow \mathbf{B} \setminus \mathbf{B}_f \subseteq \rho$. This implies the complementariness of CBC operators.

Definition 8 (Monotonicity). *A contracting operator (OC or CBC), μ, is called monotonous if $\mathbf{B} \subseteq \mathbf{B}' \Rightarrow \mu(\mathbf{B}, \rho) \subseteq \mu(\mathbf{B}', \rho)$.*

Proposition 4. *Given a set of n OC operators, $\{\mathrm{OC}_k\}$, and a set of n CBC operators, $\{\mathrm{CBC}_k\}$, which enforce the monotonicity, where OC_k and CBC_k are defined in k-dimension ($k \leq n$). The function $f : \mathbb{I}^n \times \mathcal{R}_n \to \{\mathtt{feasible}, \mathtt{infeasible}, \mathtt{unknown}\}$ defined by the following rules is a feasibility checker:*

(i) $f(\mathbf{B}, \rho) = \mathtt{infeasible} \Leftrightarrow \mathrm{OC}_k(\mathbf{B}|_{vars(\rho)}, \rho) = \emptyset$, where $k = |vars(\rho)|$
(ii) $f(\mathbf{B}, \rho) = \mathtt{feasible} \Leftrightarrow \mathrm{CBC}_k(\mathbf{B}|_{vars(\rho)}, \rho) = \emptyset$, where $k = |vars(\rho)|$

Proof of Proposition 4 is straightforward due to Proposition 1, Proposition 2 and Definition 5. Proposition 4 gives a way to construct a feasibility checker from OC and CBC operators enforcing the monotonicity. Proposition 2 and Proposition 3 imply that CBC operators can be constructed by OC operators and that *Complementary-Boxing* makes it possible to isolate completely feasible regions, $\mathbf{B} \setminus \mathrm{CBC}(\mathbf{B}, \rho)$, w.r.t. some constraints. When applying a CBC operator to a box with respect to a constraint results in an empty set, it can be deduced that the box completely satisfies that constraint. We then define a splitting operator based on Complementary-Boxes, which consists of splitting around Complementary-Boxes, to isolate the feasible regions.

Definition 9 (Box Splitting Operator: BS). *A Box splitting operator is a function $\mathrm{BS} : \mathbb{I}^n \times \mathbb{I}^n \to \mathcal{P}(\mathbb{I}^n)$, which takes as input two boxes such that the former contains the latter, splitting the outer box along some facets of the inner one. The output boxes must be disjoint.*[7]

In the algorithm we propose, *Box splitting*, which partitions a region around a Complementary-Box, is applied in combination with dichotomous splitting.

---

[7] This is a generic definition for partitioning a region around a box contained in it, given in [14].

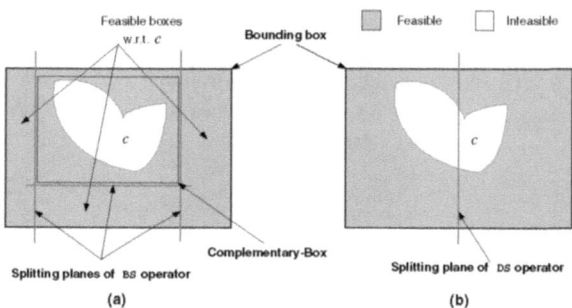

Fig. 2. (a) Box Splitting: splitting around a box (e.g. Complementary-Box); (b) Dichotomous Splitting: splitting the original domain of a variable into two halves

The latter is used either when Complementary-Boxing produces no reduction or when Box splitting results in too small boxes. Figure 2 illustrates the notion of Box splitting.

4.2   Concise Approximations near Boundaries Using EVR

Stop Contracting over Inactive Dimensions. We first observe that a better alignment of boxes near boundaries of the solution space can be obtained by finely controlling applications of contracting operators during search. More precisely, whenever a dimension, i, of a box, $\mathbf{B}$, is bounded by ε_i, one can prevent contracting operators from contracting $\mathbf{B}$ over this dimension in order to obtain better alignments and performances. Given a box, $\mathbf{B}$, a constraint set, $\mathcal{C}$, and a precision vector, ε. A dimension, i, of $\mathbf{B}$ is called an *active dimension* if $\sup(\mathbf{B}|_i) - \inf(\mathbf{B}|_i) > \varepsilon_i$ and the corresponding variable, x_i, occurs in some active constraint. Otherwise, it is called an *inactive dimension*. A contracting operator which only works on active dimensions of boxes is called a *restricted-dimensional contracting operator* [11]. We denote by $\mathtt{OC_{rd}}$ and $\mathtt{CBC_{rd}}$ the restricted-dimensional contracting operators corresponding to $\mathtt{OC}$ and $\mathtt{CBC}$ operators, respectively.

Compacting Aligned Boxes. Once a better alignment is obtained, the question is how such a set of aligned boxes can be compacted into a smaller set. We propose to use the *Extreme Vertex Representation* (EVR) of orthogonal polyhedra for that purpose. The basic idea is that the finite unions of boxes delivered by a box-covering solver define *orthogonal polyhedra* for which improved representations can be used. An orthogonal polyhedron can be naturally represented as a finite union of disjoint boxes. Such a representation is called the *Disjoint Box Representation* (DBR) in computational geometry. The EVR is a way of compacting DBR [1, 2, 3]. We now recall some basic concepts related to EVR. We refer the reader to [2, 3] for further details. The concepts are presented for a particular type of orthogonal polyhedra, called *griddy polyhedra*. A griddy

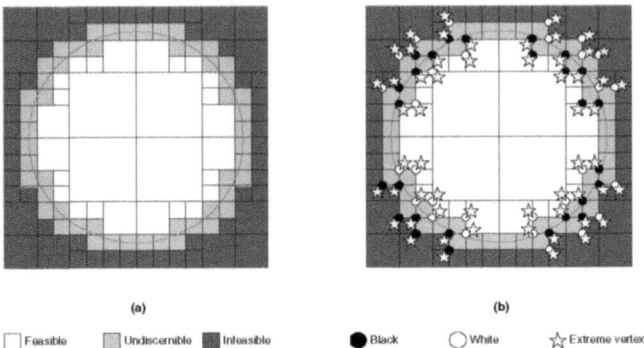

(a)   (b)

☐ Feasible   ▨ Undiscernible   ▩ Infeasible   ● Black   ○ White   ☆ Extreme vertex

Fig. 3. (a) DBR and (b) Extreme vertices (with their colors) of union approximations

polyhedron [3] is generated from unit hyper-cubes with integer-valued vertices. Since arbitrary orthogonal polyhedra can be obtained from griddy ones by a bijection between vertex indices of the former and integer-valued vertices of the later, the results on EVR are not affected by this simplification. For simplicity, polyhedra are assumed to live inside a bounded subset $\mathbf{X} = [0, m]^d \subseteq \mathbb{R}^d$ (in fact, the results will hold also for $\mathbf{X} = \mathbb{R}_+^d$). Let $\mathbf{x} = (x_1, ..., x_d)$ be a grid point of the elementary grid $\mathcal{G} = \{0, 1, ..., m-1\}^d \subseteq \mathbb{N}^d$. For every point $\mathbf{x} \in \mathbf{X}$, $\lfloor \mathbf{x} \rfloor$ is the grid point corresponding to the integral part of the components of $\mathbf{x}$. The elementary box associated with $\mathbf{x}$ is the closed subset of $\mathbf{X}$ of the form $\mathbf{B}(\mathbf{x}) = [x_1, x_1 + 1] \times ... \times [x_d, x_d + 1]$. The set of all boxes is denoted by $\mathcal{B}$. A griddy polyhedron P is a union of elementary boxes, i.e. an elementary of $2^{\mathcal{B}}$.

Definition 10 (Color Function). *Let P be a griddy polyhedron. The color function $c : \mathbf{X} \to \{0, 1\}$ is defined as follows: if $\mathbf{x}$ is a grid point then $c(\mathbf{x}) = 1 \Leftrightarrow \mathbf{B}(\mathbf{x}) \subseteq P$; otherwise, $c(\mathbf{x}) = c(\lfloor \mathbf{x} \rfloor)$.*

We say that a grid point $\mathbf{x}$ is black (respectively, white) and that $\mathbf{B}(\mathbf{x})$ is full (respectively, empty) when $c(\mathbf{x}) = 1$ (respectively 0). A *canonical representation scheme* for $2^{\mathcal{B}}$ (or $2^{\mathcal{G}}$) is a set $\mathcal{E}$ of syntactic objects such that there is some bijective function $\Psi : \mathcal{E} \to 2^{\mathcal{B}}$.

Definition 11 (Extreme Vertex). *A grid point $\mathbf{x}$ is called an extreme if $\tau(\mathbf{x}) = 1$, where $\tau(\mathbf{x})$ denotes the parity of the number of black grid points in $\mathcal{N}(\mathbf{x}) = \{x_1 - 1, x_1\} \times ... \times \{x_d - 1, x_d\}$ (the neighborhood of $\mathbf{x}$).*

Figure 3 illustrates the notion of EVR on a simple example. The fundamental theorem presented in [2, 3] shows that any griddy polyhedron can be canonically represented by the set of its extreme vertices (and their colors). The extreme vertex representation improves the space required for storing orthogonal polyhedra by an order of magnitude [1, 2, 3]. It also enables the design of efficient algorithms for fundamental operations on orthogonal polyhedra (e.g.

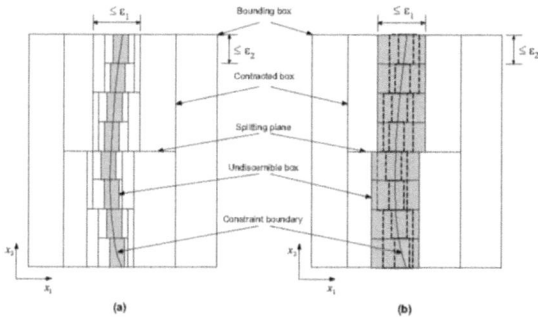

Fig. 4. Constraint boundary: (a) unaligned boxes produced by standard covering; (b) stop contracting over inactive dimensions and combine aligned boxes using EVR

membership, set-theoretic operations) [1, 2, 3]. In particular, effective transformation between DBR and EVR can be proposed for low-dimensional or small-size (i.e. m is small) polyhedra [1, 3]. For example, in three-dimension, the average experimental (time) complexity of converting an EVR to a DBR is far less than quadratic but slightly greater than linear in the number of extreme vertices [1]. Results in [3] also imply that, in fixed dimension, the time complexity of converting a DBR to an EVR using XOR operations is linear in the number of boxes in DBR. We propose to exploit these effective transformation schemes to produce a compact representation of contiguous aligned boxes using the following procedure:

1. Produce a better alignment of the boxes along the boundaries of constraints. This is done by preventing the unnecessary application of contracting operators over inactive dimensions. Figure 4 shows the better alignment produced for a set of nearly aligned boxes of an undiscernible approximation. The original set of 8 small boxes (Figure 4-a) reduces to two groups of 4 aligned boxes (Figure 4-b) without altering the predefined interval-based precision.
2. The set of aligned boxes in each group, S_1, is converted to EVR and then back to DBR to get a set of combined boxes, S_2 (containing only one box in this case). Due to the properties of EVR, S_2 is more concise than S_1. Figure 4-b shows how this conversion procedure reduces the two groups of 4 boxes to two (gray) boxes.

Such a procedure can theoretically be applied in any dimension. Due to the efficiency of EVR in low dimension, we however restrict its application to low-dimensional or small-size regions of the search space in our implementation (see Section 5).

5   Algorithms

We now present an algorithm called UCA6$^+$ (Figures 5 and 6). It takes as input a non-linear NCSP, $P = (\mathcal{V}, \mathcal{C}, \mathcal{D})$, and returns the **Union**$^\mathcal{I}$ and **Union**$^\mathcal{U}$ approximations of ρ_P. UCA6$^+$ is an extension of ICAb5 [10] and UCA6 [4] to include the application of extreme vertex representation of orthogonal polyhedra and the use of restricted-dimensional contracting operators. Hereafter, **B** denotes a bounding box of the current subproblem. Originally, this bounding box is set to $\mathcal{D}$. For convenience, we denote **Union**$^\mathcal{X}(\mathbf{B} \cap \rho_\mathcal{C})$ by **Union**$^\mathcal{X}(\mathbf{B}, \mathcal{C})$, where $\mathcal{X} \in \{\mathcal{O}, \mathcal{I}, \mathcal{U}\}$. UCA6$^+$ constructs the approximations **Union**$^\mathcal{I}(\mathbf{B}, \mathcal{C})$ and **Union**$^\mathcal{U}(\mathbf{B}, \mathcal{C})$, hence **Union**$^\mathcal{O}(\mathbf{B}, \mathcal{C})$ can be computed as the union of these two approximations.

UCA6$^+$ proceeds by recursively repeating three main steps:[8] *(i)* using OC$_{rd}$ operators to contract the current bounding box to a tighter one; *(ii)* using CBC$_{rd}$ operators to get a list of Complementary-Boxes w.r.t. each active constraint and w.r.t. the new bounding box, the constraints that make empty Complementary-Boxes are removed; finally, *(iii)* combining Dichotomous splitting (DS) with Box splitting (BS) for the whole set of active constraints. In practice, equalities usually define surfaces, we then do not need to perform step *(ii)* for such constraints (see Figure 4).

getSplit() is a function that returns the splitting mode to be used for splitting the current box. The current splitting mode can be inferred from the history of the current box (e.g. the splitting mode of the parent box). In contrast to DMBC, the DS operator used for UCA6$^+$ only tries to dichotomize over the active dimensions. This avoids splitting boxes into a huge number of tiny boxes. Moreover, in UCA6$^+$ constraints are removed gradually whenever an empty Complementary-Box is computed w.r.t. those constraints. The dimension with the greatest size is preferred for DS. For the pruning to be efficient, BS splits along some facet of a Complementary-Box only if that produces sufficiently large boxes, the Complementary-Box itself excepted. This estimation is done using a pre-determined *fragmentation ratio*. In Figure 5 and 6, $\mathcal{S}_{inn}$ and $\mathcal{S}_{und}$ are global variables denoting the set of boxes (and active constraints, if exist) of **Union**$^\mathcal{I}(\mathbf{B}_0, \mathcal{C}_0)$ and **Union**$^\mathcal{U}(\mathbf{B}_0, \mathcal{C}_0)$, respectively. We use a list, *WList*, to store the subproblems waiting to be processed.

chooseTheBest() is a function to choose the best Complementary-Box and the respective constraint based on some criteria to maximize the space surrounding the Complementary-Box. The other Complementary-Boxes can be memorized for improving the Complementary-Boxing of child search nodes.

In Figure 6, the function *solveQuickly*() constructs **Union**$^\mathcal{I}$ and **Union**$^\mathcal{U}$ approximations for low-dimensional subproblems whose bounding box has D_{stop} active dimensions at the most. The output is compacted using EVR. The use of OC$_{rd}$ and CBC$_{rd}$ operators for the purpose of narrowing produces a better alignment of boxes along the boundaries. This allows for using EVR to combine the contiguous aligned boxes. *solveQuickly*() uses a feasibility checker, called FC,

---

[8] UCA6 does the similar steps, except *solveQuickly*(), but it uses OC and CBC operators.

function $\text{UCA6}^+(\mathbf{B}_0, \mathcal{C}_0, \varepsilon, \text{FC}, \text{OC}_{\text{rd}}, \text{CBC}_{\text{rd}}, D_{stop})$

    $\mathcal{S}_{inn} := \emptyset; \ \mathcal{S}_{und} := \emptyset; \ WList := \emptyset;$ /* $\mathcal{S}_{inn}, \mathcal{S}_{und}$ are global lists to be return*/

    if $solveQuickly(\mathbf{B}_0, \mathcal{C}_0, \varepsilon, \text{FC}, \text{OC}_{\text{rd}}, \text{CBC}_{\text{rd}}, WList, D_{stop})$ **then return;**

    while $WList \neq \emptyset$ **do**   /* Waiting list of subproblems is not empty */

        $(\mathbf{B}, \mathcal{C}, \{\mathbf{CB'}_c\}) := get(WList);$ /* set $\{\mathbf{CB'}_c\}$ was optionally memorized*/

        for each $c \in \mathcal{C}'$ **do**   /* $\mathcal{C}' \subseteq \mathcal{C}$, it is set to $\mathcal{C}$ or dynamically computed based on $\{\mathbf{CB'}_c\}$ */

            $\mathbf{CB}_c := \text{CBC}_{\text{rd}}(\mathbf{B}, c) \text{ or } \text{CBC}_{\text{rd}}(\mathbf{B} \cap \mathbf{CB'}_c, c) \text{ or } \mathbf{B} \cap \mathbf{CB'}_c;$ /* depends on $\mathcal{C}'$ */

            if $\mathbf{CB}_c = \emptyset$ **then** $\mathcal{C} := \mathcal{C} \setminus \{c\};$ /* c is redundant in $\mathbf{B}$ (Proposition 2) */

            if $\mathbf{CB}_c = \emptyset$ **or** $\mathbf{CB}_c = \mathbf{B}$ **then** $\mathcal{C}' := \mathcal{C}' \setminus \{c\};$

        end-for

        if $\mathcal{C} = \emptyset$ **then**

            $store(\mathbf{S}_{inn} \leftarrow \mathbf{B});$ /* No active constraint, $\mathbf{B}$ is feasible */

            continue while;   /* do the next loop of **while** */

        end-if

        $Splitter := getSplit();$ /* Get a splitting mode, heuristics can be used */

        if $Splitter = \text{BS}$ **then**   /* The splitting mode is Box Splitting */

            $\mathbf{CB}_b := chooseTheBest(\{\mathbf{CB}_c \mid c \in \mathcal{C}'\});$ /* e.g. choose the smallest box */

            $(\mathbf{B}_1, \ldots, \mathbf{B}_k) := \text{BS}(\mathbf{B}, \mathbf{CB}_b);$ /* Box Splitting: failed or $\exists \mathbf{B}_i \supseteq \mathbf{CB}_b$ */

            if $\mathcal{C}' = \emptyset$ **or** BS *failed* **then** $Splitter := \text{DS};$

        end-if

        if $Splitter = \text{DS}$ **then** $(\mathbf{B}_1, \ldots, \mathbf{B}_k) := \text{DS}(\mathbf{B});$ /* Dichotomize $\mathbf{B}$ into disjoint boxes */

        for $i = 1$ **to** k **do**

            $\mathcal{C}_i := \mathcal{C}; \ \mathcal{C}'_i := \mathcal{C}';$

            if $Splitter = \text{BS}$ **and** $\mathbf{B}_i \cap \mathbf{CB}_b = \emptyset$ **then**

                $\mathcal{C}_i := \mathcal{C}_i \setminus \{b\}; \ \mathcal{C}'_i := \mathcal{C}'_i \setminus \{b\}$ /* b is redundant (Complementarity of CBC) */

                if $\mathcal{C}_i = \emptyset$ **then**

                    $store(\mathbf{S}_{inn} \leftarrow \mathbf{B}_i);$ /* No active constraint, $\mathbf{B}_i$ is feasible */

                    continue for;   /* do the next loop of **for** */

                end-if

            end-if

            $solvable := solveQuickly(\mathbf{B}_i, \mathcal{C}_i, \varepsilon, \text{FC}, \text{OC}_{\text{rd}}, \text{CBC}_{\text{rd}}, WList, D_{stop});$

            if not $solvable$ **then** $memorize(WList \leftarrow \{\mathbf{CB}_c \mid c \in \mathcal{C}'_i\});$ /* This is optional */

        end-for

    end-while

end   /* of UCA6^+ */

Fig. 5. The algorithm UCA6^+

function $solveQuickly(\mathbf{B}, \mathcal{C}, \varepsilon, \mathrm{FC}, \mathrm{OC_{rd}}, \mathrm{CBC_{rd}}, WList, D_{stop})$
    $\mathbf{B}' := \mathrm{OC_{rd}}(\mathbf{B}, \mathcal{C});$
    if $\mathbf{B}' = \emptyset$ **then return** TRUE;   /* B is infeasible, the problem has been solved */
    if $isEpsilonBox(\mathbf{B}', \mathcal{C}, \varepsilon, \mathrm{FC})$ **then return** TRUE;   /* The problem has been
solved */
        if $\mathbf{B}'$ *has at most* D_{stop} *active dimensions* **then**   /* Resort to another technique
*/

            $< \mathcal{S}'_{inn}, \mathcal{S}'_{und} > := DimStopSolver(\mathbf{B}', \mathcal{C}, \varepsilon, \mathrm{FC}, \mathrm{OC_{rd}}, \mathrm{CBC_{rd}});$
            /* *combine*() does the conversions DBR $\rightarrow$ EVR $\rightarrow$ DBR in D_{stop}-dimension */
            $store(\mathcal{S}_{inn} \leftarrow combine(\mathcal{S}'_{inn}));$ /* Store in the global list of feasible boxes */
            $store(\mathcal{S}_{und} \leftarrow combine(\mathcal{S}'_{und}));$ /* Store in the global list of undiscernible boxes
*/

        return TRUE;   /* The problem has been solved */
    end-if
    $put(WList \leftarrow (\mathbf{B}', \mathcal{C}));$ /* put the subproblem into the waiting list */
    return FALSE;   /* The problem has not been solved yet */
end   /* of *solveQuickly* */

function $isEpsilonBox(\mathbf{B}, \mathcal{C}, \varepsilon, \mathrm{FC})$
    if $\mathbf{B}$ *has no active dimension* **then**   /* B is an ε-bounded box in the variable space
of $\mathcal{C}$ */
        switch $\mathrm{FC}(\mathbf{B}, \rho_\mathcal{C})$:   /* Identify the feasibility of B w.r.t. $\mathcal{C}$ */
            case feasible : $store(\mathcal{S}_{inn} \leftarrow \mathbf{B});$ /* B is feasible, store it */
            case unknown : $store(\mathcal{S}_{und} \leftarrow (\mathbf{B}, \mathcal{C}));$ /* B is undiscernible, store it */
        end-switch
        return TRUE;
    end-if
    return FALSE;
end   /* of *isEpsilonBox* */

Fig. 6. The function *solveQuickly*()

to check if an ε-bounded box is `feasible`, `infeasible` or `unknown` (then called
undiscernible). Though the FC in our implementation uses OC and CBC operators
for checking the feasibility of ε-bounded boxes, it is however not restricted to
a specific feasibility checker.

For efficiency, *solveQuickly*() allows resorting to a secondary search technique, *DimStopSolver*(), to solve the low-dimensional subproblems whose
bounding box has at most D_{stop} active dimensions. Good candidates for
small D_{stop} can be either the 2^k-tree based solver presented in [8] or a simple
grid-based solver[9]. Variants of DMBC or UCA6 using the *restricted-dimensional*
contracting operators can alternatively be used. For a given subproblem,
DimStopSolver() constructs the sets $\mathcal{S}'_{inn}$ and $\mathcal{S}'_{und}$ which are **Union**$^{\mathcal{I}}$ and
Union$^{\mathcal{U}}$ of the subproblem, respectively. These two sets are represented in DBR.

---

[9] A simple grid-based solver splits variable domains into a grid and solves the problem
in each grid element.

They are converted to EVR and then back to DBR to combine each group of contiguous aligned boxes into a bigger equivalent box. This operation is represented by the function *combine*() in Figure 6.

Proposition 5. *Given a feasibility checker,* FC. *The algorithm* UCA6$^+$ *computes the union approximations and has a predefined interval-based precision* ε *w.r.t.* FC.

Sketch of Proof: The conclusion can be deduced, informally, from the observations: *(i)* **Union**$^{\mathcal{I}}$ and **Union**$^{\mathcal{U}}$ are disjoint, **Union**$^{\mathcal{O}}$ = **Union**$^{\mathcal{I}}$ ∪ **Union**$^{\mathcal{U}}$; *(ii)* No solution is lost due to the completeness of OC operator; *(iii)* All the inner boxes (i.e. the boxes in **Union**$^{\mathcal{I}}$) are sound due to Definition 7 and Proposition 2; *(iv)* **Union**$^{\mathcal{U}}$ is equivalent to a union of the boxes (before applying EVR-DBR conversions) which have no active dimension (see the function *isEpsilonBox*() in Figure 6) and cannot be classified as feasible or infeasible using the feasibility checker FC. That is due to the properties of EVR-DBR conversion and due to the fact that each box which has no active dimension has the same feasibility (under feasibility checkers) with its projection, an ε-bounded box, on the space defined by all the variables in the active constraints.

6   Preliminary Experiments

We now present a preliminary evaluation on the following small set of typical problems (with different properties of constraints and solution space).

CD (Column Design), **FD** (Fatigue Design) and **TD** (Truss Design) are three engineering design examples. Their complete descriptions are available at http://imacsg4.epfl.ch:8080/PGSL/ and http://liawww.epfl.ch/Coconut-benchs/. In Table 1, the considered instance of **CD** is the one that finds $(a, b, e) \in [0.01, 2] \times [0.01, 1] \times [0.05, 0.1]$ given that $P = 400kN$, $H = 6m$ and $L = 1m$, where a and b are in meter, e is in decimeter. The **FD** instance considered is the one that finds $(L, qf, Z) \in [10, 30] \times [70, 90] \times [0.1, 10]$ for a given number of years to fatigue failure $years = 100$, where Z is scaled up 100 times in unit. The considered instance of **TD** is a simplified one that finds $x_1, y_1 \in [0.01, 10]$. **WP** and **P3** were described in Section 2. **P2** = $\{x^2 \leq y,\ \ln y + 1 \geq z,\ xz \leq 1,\ x \in [0, 15],\ y \in [1, 200],\ z \in [-10, 10]\}$.

For evaluation purposes, we have implemented the algorithms DMBC, UCA6, UCA6$^+$ using the same data structure and the same standard contracting operators. We have also implemented a direct transposition, called UCA5, of ICAb5 in [10] to solving NCSPs, and a version of DMBC including the negation test. This enhanced DMBC, called DMBC$^+$, can therefore check whether a box is completely feasible or not. Our experiments discarded DMBC as a reasonable candidate for this kind of problems. It usually produces a huge number of boxes, each of which is ε-bounded.

The tests shown in Table 1 ran with *fragmentation ratio* = 0.25, $D_{stop} = 1$, and FC given by Proposition 4 (with the precision set to 1). The OC operator

Table 1. Typical test results

Prob.	ε	DMBC+ (DS), ¬MEM	UCA6 (DS), MEM	UCA6+ (DS), MEM	UCA5 (BS+DS), ¬MEM	UCA6 (BS+DS), MEM	UCA6+ (BS+DS), ¬MEM
WP	0.1	22.07s / 0.992 2753 / 2620	6.73s / 0.992 2753 / 2620	5.98s / 0.991 2489 / 2147	5.11s / 0.994 1738 / 2788	4.77s / 0.994 1573 / 2791	3.97s / 0.993 1176 / 1585
TD	0.01	81.53s / 0.997 3900 / 2917	26.45s / 0.997 3900 / 2917	26.01s / 0.997 3895 / 1970	14.96s / 0.999 2832 / 3270	13.43s / 0.999 1313 / 3496	3.59s / 0.998 53 / 50
P3	0.1	>10h / n/a >110000 / 150000	615.98s / 0.907 33398 / 38006	530.16s / 0.912 30418 / 28229	87.09s / 0.980 10784 / 29888	135.28s / 0.980 12113 / 38808	1.41s / 0.919 406 / 970
P2	0.1	>10h / n/a >120000 / 180000	4959.76s / 0.973 108701 / 100027	4433.06s / 0.974 106320 / 78755	281.51s / 0.995 21872 / 55901	293.09s / 0.995 18722 / 55063	7.32s / 0.975 1873 / 3225
FD	0.1	3878.47s / 0.987 42178 / 66940	1278.82s / 0.987 42178 / 66940	740.05s / 0.984 42084 / 47138	394.29s / 0.992 51882 / 65536	439.07s / 0.992 26378 / 70170	211.91s / 0.986 10341 / 35126
CD	0.01	1308.63s / 0.638 8957 / 22132	468.21s / 0.638 8873 / 21974	389.12s / 0.619 8354 / 14857	538.21s / 0.830 12729 / 25079	493.57s / 0.828 9922 / 25344	352.96s / 0.616 2826 / 12922

was implemented with *IloGenerateBounds* in ILOG Solver 5.2 [6]. The precision of the contracting operators used for narrowing bounding boxes was set to 1. Let ε be the interval-based precision for the algorithms. The secondary search technique used for UCA6+ is a simple grid-based one. The terms (DS) and (BS+DS) indicate the splitting strategies enforced upon the algorithms, MEM means the memorization of Complementary-Boxes. Each cell in the table has two rows. The first shows time and ratio of inner volume to total volume, the second the number of boxes in **Union**$^{\mathcal{I}}$ and **Union**$^{\mathcal{U}}$, respectively.

Other tests on tens of similar problems show that the best gains, in running time and number of boxes, of the algorithms UCA5, UCA6 and UCA6+ over DMBC+ are obtained for problems with low-arity constraints (w.r.t. arity of problems, e.g. **P2**, **P3**). In all the tests, UCA6+ (BS+DS) with either MEM or ¬MEM is better than the other algorithms in running time and number of boxes. The best gains

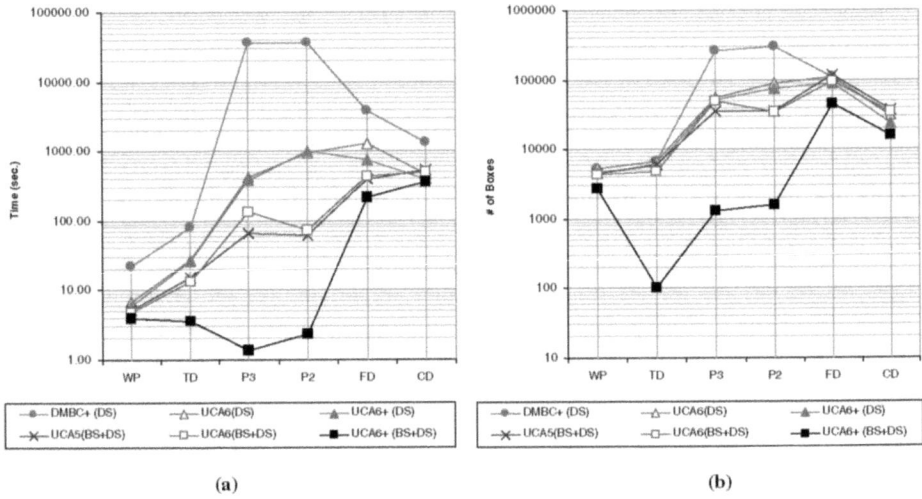

(a)   (b)

Fig. 7. Logarithmic charts for (a) running time, and (b) total number of boxes

are obtained when the non-linear constraints contain some nearly axis-parallel regions (e.g. **P2**, **P3**, **TD**). $UCA6^+$ ($BS+DS$) is slightly less accurate than $UCA5$ and $UCA6$ in volume measure, but the situation is improved when ε is reduced (i.e. in higher precision) and it is not important in real applications.

The preliminary experiments are therefore encouraging enough to warrant further investigations and in-depth evaluations based on other combinations of the control parameters and higher values of D_{stop} in combination with variants of 2^k-tree solvers.

7 Conclusion

Interval-constraint based solvers are usually designed to deliver point-wise solutions. They are consequently inadequate for solving numerical problems with non-isolated solutions. In this paper, we propose a box-covering technique for computing inner and outer approximations of NCSPs that remedies this state of affairs. The approach works for general non-linear NCSPs with mixed equality/inequality constraints, especially for inequalities. It combines the compactness of extreme vertex representation of orthogonal polyhedra with an adapted splitting policy. This allows for gains in performance and space requirements. The quality of the output is also enhanced by *guaranteed feasible boxes* when solution space is solid. In practice, NCSPs with non-isolated solutions often occur as subproblems of higher-dimensional problems. For such purposes, the feasibility checker can be relaxed or run with low precision, running time will hence be much improved. In future work, we therefore plan to investigate collaboration strategies between our techniques and standard interval-based solvers and plan to do more experiments with higher values of D_{stop} in combination with variants of 2^k-tree solvers.

Acknowledgments

We would like to thank Prof. Boi Faltings and Prof. Arnold Neumaier for their valuable discussions. Support for this research was provided by the European Commission and the Swiss Federal Education and Science Office (OFES) through the COCONUT project (IST-2000-26063).

References

[1] Aguilera, A.: Orthogonal Polyhedra: Study and Application. PhD thesis, Universitat Politècnica de Catalunya, Barcelona, Spain (1998)

[2] Bournez, O., Maler, O., Pnueli, A.: Orthogonal Polyhedra: Representation and Computation. F. Vaandrager and J. Van Schuppen (Eds.), Hybrid Systems: Computation and Control, LNCS 1569 (1999) 46–60 Springer.

[3] Bournez, O., Maler, O.: On the Representation of Timed Polyhedra. In: Proceedings of International Colloquium on Automata Languages and Programming (ICALP'2000). (2000)

[4] Silaghi, M.C., Sam-Haroud, D., Faltings, B.: Search Techniques for Non-linear CSPs with Inequalities. In: Proceedings of the 14th Canadian Conference on AI. (2001)

[5] Van Hentenryck, P.: A Gentle Introduction to Numerica. (1998)

[6] ILOG: ILOG Solver. Reference Manual. (2002)

[7] Jaulin, L.: Solution Globale et Guarantie de Problèmes Ensemblistes: Application à l'Estimation Non Linéaire et à la Commande Robuste. PhD thesis, Université Paris-Sud, Orsay (1994)

[8] Sam-Haroud, D., Faltings, B.: Consistency Techniques for Continuous Constraints. Constraints **1(1-2)** (1996) 85–118

[9] Garloff, J., Graf, B.: Solving Strict Polynomial Inequalities by Bernstein Expansion. In: Symbolic Methods in Control System Analysis and Design. (1999) 339–352

[10] Benhamou, F., Goualard, F.: Universally Quantified Interval Constraints. In: Proceedings of the 6th International Conference on Principles and Practice of Constraint Programming (CP'2000). (2000) 67–82

[11] Vu, X.H., Sam-Haroud, D., Silaghi, M.C.: Approximation Techniques for Nonlinear Problems with Continuum of Solutions. In: Proceedings of The 5th International Symposium on Abstraction, Reformulation and Approximation (SARA'2002), Canada (2002) 224–241

[12] Lhomme, O.: Consistency Techniques for Numeric CSPs. In: Proceedings of IJCAI'93. (1993)

[13] Collavizza, H., Delobel, F., Rueher, M.: Extending Consistent Domains of Numeric CSP. In: Proceedings of IJCAI'99. (1999)

[14] Van Iwaarden, R.J.: An Improved Unconstrainted Global Optimization Algorithm. PhD thesis, University of Colorado at Denver, USA (1996)

Benchmarking Global Optimization and Constraint Satisfaction Codes

Oleg Shcherbina[1], Arnold Neumaier[1], Djamila Sam-Haroud[2], Xuan-Ha Vu[2],
and Tuan-Viet Nguyen[2]

[1] University of Vienna, Austria
[2] Swiss Federal Institute of Technology, Lausanne, Switzerland

Abstract. A benchmarking suite describing over 1000 optimization problems and constraint satisfaction problems covering problems from different traditions is described, annotated with best known solutions, and accompanied by recommended benchmarking protocols for comparing test results.

1 Introduction

Global optimization problems and constraint satisfaction problems are NP-hard, and a widely held conjecture states that no polynomial, i.e., universally fast, algorithms solving such problems exist.

It is thus necessary to assemble a sufficient set of relevant and well-categorized problems in order to be able to evaluate experimentally different approaches, techniques, and/or solution strategies, and to compare them according to various performance measures.

The objective of the current work is to provide a **benchmark** consisting of a comprehensive suite of representative problems, covering as far as possible all the categories of problems that are relevant either from a scientific, technical, or industrial point of view.

A good benchmark must be one that can be interfaced with our framework and with other, existing systems, in a way that a sufficient number of comparative results can be obtained. There are various smaller-scale benchmark projects for partial domains, in particular the benchmarks for local optimization by MITTELMANN [11]. A very recent web site by GAMS World [6] started collecting real life global optimization problems with industrial relevance, but currently most problems on this site are without computational results. Our benchmark includes a large part of the problems from these two projects.

All problems in our benchmark are represented in a common format suitable for automatic execution on global optimization and constraint satisfaction software. To ensure that benchmarking results are comparable across different solvers and environments, a **benchmarking protocol** is defined, whose execution on the benchmark will provide a number of objective performance measures for any implementation of global optimization and constraint satisfaction techniques.

C. Bliek et al. (Eds.): COCOS 2002, LNCS 2861, pp. 211–222, 2003.

With the present benchmarking suite, it is the first time that a large benchmark

- is made publicly available
- in a uniform, widely accessible format,
- covering problems from different traditions – nonlinear programming, global optimization, and constraint satisfaction,
- including most problems from the more restricted traditional collections of benchmarking problems as particular cases,
- checked for consistency,
- annotated with type and complexity information,
- (almost) complete with best known solutions, and
- accompanied by benchmarking protocols for comparing test results.

2   Description of the Benchmarking Suite

The delivered benchmarking suite is a comprehensive collection of 329 constraint satisfaction and 993 optimization problems from academic, industrial and real-life environments. Executable versions of these test problems, as well as information on their sources, are publicly available at

> http://www.mat.univie.ac.at/~neum/glopt/coconut/benchmark.html

The problems range in difficulty from easy to very challenging, their sizes from a few number of variables to over 1000 variables. These test problems come from both the research literature and a wide spectrum of applications, including:

- Chemical engineering (pooling and blending, separation, heat exchanger network design, phase and chemical equilibrium, reactor network synthesis, etc.)
- Computational chemistry (including molecular design)
- Civil engineering problems
- Robotics problems
- Operations research problems
- Economics problems (including Nash equilibrium, Walrasian equilibrium, and traffic assignment problems)
- Multicommodity network flow problems
- Process design problems
- Stability analysis problems
- VLSI chip design problems
- Portfolio optimization problems

Older Collections Covered. The problem suite incorporates as integral part most problems from the CUTE test collection [8] (covering among others the Argonne test set, the Hock and Schittkowski collection, the Dembo network problems, the Gould quadratic programs, etc.), from the handbook of test problems in local and global optimization [2], from the GLOBAL Library [6], and from the Numerica [15] test problem collection; in addition many other constraint satisfaction problems from the literature.

Some problems (e.g. linear or convex quadratic programs) are known to be solvable to global optimality by local optimization alone. They were retained in the benchmark to be able to measure the overhead which global solvers have in order to prove optimality, compared with local solvers which are usually significantly faster and find of course for convex problems a global minimum.

Common Format. All test problems are coded in the AMPL modeling language. AMPL [3] is a flexible and convenient algebraic modeling language enabling rapid prototyping and model development. It is of widespread use in the optimization community, as attested by the large number of existing interfaces with state-of-the-art optimization solvers http://www.ampl.com/solvers.html. Unfortunately, no current modeling system allows the input of interval data reflecting uncertainties in the parameters specified. Such a facility would be important for fully rigorous search.

The CUTE problems existed already in AMPL format, coded by Hande Y. Benson, and made available in the collection of AMPL files by VANDERBEI [14]. All other problems were either newly coded in AMPL, or automatically translated from existing GAMS models using GMS2XX [7] and GAMS/Convert [4].

As far as reasonable, all problems were checked for correctness; inconsistencies with information available from other sources were removed if possible. Information about (approximate) solutions and putative global minimizers and minima were provided in all but a few cases (where runtime constraints became active).

A few problems from the collections mentioned are missing in the present benchmark because of our inability to get a valid authentic version (e.g., the 'handbook' [2] contains numerous inconsistencies), or because of the presence of constructs or functions not supported by our current system (such as if, erf).

The AMPL code for all problems in the benchmark is available online at the web site mentioned, through links from tables with one line summaries of problem characteristics.

Typology of Problems. The main criteria for categorizing the test problems are mathematical properties reflecting their degree of formal complexity. This is not equivalent with computational complexity, but typically more complex objectives and constraints require more complex algorithms to handle them.

A second criterion is the problems closeness to applicability in real-life applications.

Each problem is classified as in the CUTE collection of test problems, except that problems with a V code for the number of variables or constraints have the default value for that number after the V. In addition, a few more types of constraints are distinguished, and an optional type characterizing the solution set is provided.

The classification is a string of the form OCSD-KIT-n-m in which the letters are replaced by appropriate codes, as follows:

```
O = objective code:
        C = Constant
        L = Linear
        N = No objective
        Q = Quadratic
        S = Sum of squares
        O = Other (none of the above)
C = constraint type:
        B = Bounds on variables
        L = Linear
        N = linear Network
        Q = Quadratic
        U = Unconstrained
        X = only fiXed variables
        P = Polynomial
        T = Trigonometric
        O = Other (none of the above)
S = smoothness:
        R = Twice continuously differentiable
        I = Other
D = degree of differentiability:  0, 1, or 2

K = kind of the problem:
        A~= Academic
        M = Modeling
        R = Real application
I = internal variables:
        Y = yes, problems has useful internal variables
        N = no useful internal variables
T = Type of solution set:
        I = Isolated
        N = Non-isolated
        U = Unknown

n = number of variables, or: V = varies, followed by default
m = number of constraints, or: V = varies, followed by default
```

In most cases, we also report the number of nonzeros and nonlinear nonzeros in an internal representation used in the GAMS system [5]. These give additional complexity information.

Solutions. For all problems with nonconstant objective function we provide on the WWW site the function value of the best point found, in many cases the global optimum.

Many test problems in the current benchmarking suite contain floating-point constants. Unfortunately, the AMPL software currently does not allow to control the rounding errors in the conversion to an internal representations. For this reason, the solutions reported in the current benchmarking suite are approximate only. Usually the solutions should be affected only in insignificant decimal places but there may be a nontrivial effect in case of degeneracies. This must be kept in mind when comparing our solution information to the literature, or to rigorous solvers which have a mathematically rigorous input/output interface.

The information about the solutions, their status (feasible/local minimum/global minimum) and accuracy (approximate/verified) will be updated as we run the benchmark with verifying global solvers.

3 Benchmarking Protocol

The benchmarking suite is designed to allow researchers convenient testing of their own algorithms, and to post the results for comparison with other algorithms.

The benchmarking protocol defines the experimentation procedure and the criteria used to measure *efficiency* of the algorithm. It can be carried out within a limited amount of work, and hence be checked in regular intervals. We decided to create a benchmarking protocol that works on the full set of problems, while allowing code developers to assess progress without endless testing.

A well-designed benchmarking protocol must be able to assess work in progress as well as the final results obtained. Since it may be impractical to frequently repeat testing on problems of realistic size, the **benchmarking protocol for assessing work in progress** is designed to be executable in a limited amount of time (approximately 24h clock time on a fast computer), and gives information about successes and weaknesses to guide the further development. The **benchmarking protocol for assessing a release of a code** has all the time limits specified below multiplied by a factor of 10.

In order to compare benchmark results across different platforms, we specify that each benchmark involves the computation of a **standard unit time**, as suggested in the first global optimization benchmark by DIXON & SZEGÖ [1] in 1974. To make it unambiguous and large enough for today's computers, we define it as the CPU time needed to carry out the C++ program defined in the appendix, compiled without optimization or debugging option. It evaluates the 4-dimensional Shekel5 test function at 10^8 specified points.

Timeout limits and runtimes are given in multiples of standard unit times. We are aware of the fact that items not accounted for in our standard unit time affect the performance of global optimization methods. The above choice was made on the assumption that people who want to switch from local to global optimization are most likely to compare against performance in terms of equivalent function values.

For branch and bound codes, time spent on memory management is relevant also. Since people solving optimization problems locally only or using heuristics like simulated annealing, use floating points only, and use little memory, total runtime (and not number of function values or number of iterations) seems to be the most relevant unit to compare with – it tells how much slower or faster the complete solver is compared with heuristics.

In the following,we describe the information used and produced by the protocol (parameters, parameter settings, stopping rules and output).

Parameters and Rules The parameters and rules used by the benchmarking protocols are defined as follows:

- The two main query options defined are: running *incomplete* and *complete* search; each search option may be executed in either *approximate* or *rigorous* verification mode, on any of four *complexity classes*. This gives $2 * 2 * 4 = 16$ cases to consider.
  - A successful *incomplete search* shall mean:
    * (for problems with objective function) running the search until the best known function value was found for the first time to within a relative accuracy of 10^{-6} or an absolute accuracy of 10^{-9}, whichever is more generous;
    * (for problems without objective function) running the search until the first feasible point is found to within a relative accuracy of 10^{-3} or an absolute accuracy of 10^{-6}, whichever is more generous.
  - A successful *complete search* shall mean:
    * (for problems with objective function) locating all global optimizers, or asserting correctly that the objective function is unbounded below on the search space.
    * (for problems without objective function and a discrete solution set) finding all feasible points, or asserting correctly that none exists.
    * (for problems without objective function and nonisolated solutions) finding an explicit description of the feasible point set within a relative accuracy of at most 5 percent of the maximal side of the interval hull of the feasible point set.
  - in *approximate verification mode,*
    * the solution(s) must satisfy all constraints to within a relative accuracy of approximately 10^{-3} or an absolute accuracy of approximately 10^{-6}, whichever is more generous.
  - in *rigorous verification mode,*
    * the solution(s) must be enclosed in boxes of a relative accuracy of approximately 10^{-3} or an absolute accuracy of approximately 10^{-6}, whichever is more generous, and any such box is guaranteed with certainty to contain a feasible point, in spite of rounding errors made (but there may be additional boxes neither excluded nor guaranteed to contain a solution),
    * (for problems with objective function) the global minimum value is to be enclosed with an approximate relative accuracy of 10^{-6} or an absolute accuracy of 10^{-9}, whichever is more generous.

- All problems are assigned to one of four complexity classes according to the number of variables defining the problem:
  * *size* 1, with $0 - 9$ variables,
  * *size* 2, with $10 - 99$ variables,
  * *size* 3, with $100 - 999$ variables,
  * *size* 4, with ≥ 1000 variables.
- The problems are sorted into three libraries, two with optimization problems, one with constraint satisfaction problems.

  For testing, the problems of each library are arranged in a fixed order of increasing dimension within each complexity class. This order will probably change with time, to reflect the currently unknown real difficulty of the test problems. Thus the file defining the ordering contains a version letter which should be quoted when publishing results using the benchmark. Version A of the required ordering is specified in

  http://www.mat.univie.ac.at/~neum/glopt/coconut/probclasses.txt

  This gives a total of $3 * 4 = 12$ lists of problems, to be executed with up to four possible pairs of query options.

In some cases it is unreasonable that the full goal is achieved (for ill-conditioned solutions, or for non-isolated solution sets); in these cases, suitably relaxed goals may be specified in comments to the solution statistic provided.

- Each problem of a given complexity class gets its own timeout limits for both complete and incomplete search; and each complexity class within each of the three libraries gets total timeout limits for both complete and incomplete search.
- Part of the documentation of running the benchmark should contain:
  - the program name and version,
  - if applicable, the compiler used for creating the executable,
  - the standard unit time (and optionally other timing information),
  - the tuning parameter settings in the algorithm used,
  - starting points used (if there is a choice in the algorithm), and
  - the stopping rule used.

  In particular, for each particular benchmark run, all problems should be handled by the same set of tuning parameters and stopping rules. If only part of the benchmark is tested, the reasons should be given.
- For each query option and each tested problem, the following information (if available) should be reported:
  - success or failure with respect to
    * timeout limit reached,
    * incorrect solution(s) found, or
    * best known function value not reached;
  - function value reached (for problems with objective function).

  If a better function value than the best known one is found, this should be mentioned and the value and the coordinates of the best point found recorded; the improved values will be used in later versions of the benchmark.

- solution(s) found
- time needed for search
- accuracy reached at timeout (for first solution)
- number of nodes generated in the search
- size of remaining search space at timeout. This size is given by the residual dimension d, the residual size s, the interval hull, and the 10-logarithm of the d-volume of the interval hull of the remaining search space. (d is defined as the residual dimension of the hull of the remaining search space, and s is defined as the 10-logarithm of the sum of d-volumes of the boxes remaining.)

(The residual dimension of a box is the number d of component intervals of positive width. The d-volume is the product of these widths.)

The following global statistics should be reported for each query option and each complexity class:

- number of problems correctly solved (best known function value reached or improved/correct solution(s) identified)
- number of timeout failures (incomplete search space/first solution not found/ best known function not reached at timeout)
- number of problems incorrectly solved (for problems without objective functions)
- mean of all residual dimensions
- mean of all residual sizes.

- For problems without objective functions and with non-isolated solutions, it makes no sense to individually enumerate all (uncountably many) solutions; instead the following alternative output should be computed:
  - inner and outer approximations of the solution sets as lists of boxes.

In this case, the quality of the approximations will be reported by the volumes of the inner and the outer approximations, their quotient (if an inner approximation has nonzero volume), and by the space requirements needed for storing all the boxes. (The volume quotient will often be zero, namely if the solution set has measure zero.)

For solvers which do not provide all the output requested, the available subset of the requested output should be provided, and the limitations should be explicitly mentioned.

The benchmarking protocols are fully defined by the preceding, including the reported criteria, except for the timeout limits.

The **benchmarking protocol for assessing work in progress** is designed to be executable on the total benchmarking suite in a limited amount of time. It is implemented by setting the timeout limits to the values given in the table below, in standard time units (stu = between about 1 and 7 minutes on our machines). The timeout limit for each set is set to one tenth of the sum of the individual timeout limits.

Number of variables		1 − 9	10 − 99	100 − 999	≥ 1000	total
timeout(stu)/problem		2	10	20	40	−
Library 1	# problems	84	90	44	48	266
	timeout(stu)/set	16.8	90	88	192	386.8
Library 2	# problems	347	100	93	187	727
	timeout(stu)/set	69.4	100	186	748	1103.4
Library 3	# problems	225	76	22	6	329
	timeout(stu)/set	45.0	76	44	24	189.0
total	# problems	656	266	159	241	1322
	timeout(stu)/set	131.2	266	318	964	1679.2

This gives – for each choice of options – a total running time of about a week on our slowest (1stu=431s), and of about 23 hours on our fastest (1stu=53s) computer. Clearly, one can expect to solve only a limited number of problems within these tight time limits, especially at larger dimensions.

The **benchmarking protocol for assessing a release of a code** is implemented by running the benchmarking suite twice, setting the timeout limits first to the values in the table, and then to these values multiplied by a factor of 10.

Note: This is Version A of the benchmarking protocol. Later versions of the protocol may contain minor or major changes to the above setting, reflecting experience gained with using the protocol on various solvers and platforms.

4   Testing the Performance of Current Solvers

The solvers we started to test are MINOS [10] (a local solver, for comparison only), LGO [12], BARON [13], Numerica [15], and GlobSol [9]. Further solvers we hope to test if time permits are four recent new developments Premium Interval, LINDO global, αBB, GloptiPoly, and OptQuest.

The tables below summarize some of the main properties of these solvers, as far as known to us. Missing information is indicated by a question mark, and partial applicability by a + or − in parentheses; the dominant technique (if any) exploited by the solver is denoted by ++.

Solver	Minos	LGO	BARON	Numerica	GlobSol
Access language	GAMS	C	GAMS	Numerica	Fortran90
Integer constraints	−	+	+	−	−
search bounds	−	required	recommended	−	required
black box eval.	+	+	−	−	−
complete	−	(−)	+	+	+
rigorous	−	−	−	+	+
local	++	+	+	(+)	(+)
CP	−	−	+	++	+
other interval	−	−	−	+	++
convex	−	−	++	−	−
dual	+	−	+	−	−
available	+	+	+	+	+
free	−	−	−	(−)	+

The first two rows give the name of the solvers and the access language used to pass the problem description. The next two rows indicate whether it is possible to specify integer constraints (although the benchmark does not test this feature), and whether it is necessary to specify a finite search box within which all functions can be evaluated without floating point exceptions.

The next three rows indicate whether black box function evaluation is supported, whether the search is complete (i.e., is claimed to cover the whole search region if the arithmetic is exact and sufficiently fast) or even rigorous (i.e., the results are claimed to be valid with mathematical certainty even in the presence of rounding errors). Note that general theorems forbid a complete finite search if black box functions are part of the problem formulation, and that a rigorous search is necessarily complete.

Solver	Premium Interval	LINDO Global	αBB	GloptiPoly	OptQuest
Access language	Visual Basic	LINGO	MINOPT	Matlab	Visual Basic
Integer constraints	+	+	+	+	+
search bounds	+	?	?	−	+
black box eval.	−	−	−	−	+
complete	+	+	+	+	−
rigorous	(+)	−	−	−	−
local	+	+	+	−	+
CP	+	+	−	−	−
interval	++	+	+	−	−
convex	+	++	++	+	−
dual	−	+	−	++	−
available	+	+	−	+	+
free	−	−	−	+	−

Five further rows indicate the mathematical techniques used to do the global search. We report whether local optimization techniques, constraint propagation, other interval techniques, convex analysis, or dual (multiplier) techniques are part of the toolkit of the solver.

The final two rows indicate whether the code is available, and whether it is free (in the public domain).

References

[1] L. C. W. Dixon and G. P. Szegö, The global optimization problem: an introduction, pp. 1–15 in: L. C. W. Dixon and G. P. Szegö (eds.), Towards Global Optimisation 2, North-Holland, Amsterdam 1978.

[2] C. A. Floudas, P. M. Pardalos, C. S. Adjiman, W. R. Esposito, Z. H. Gümüs, S. T. Harding, J. L. Klepeis, C. A. Meyer and C. A. Schweiger, Handbook of Test Problems in Local and Global Optimization, Kluwer, Dordrecht 1999.

[3] R. Fourer, D. M. Gay and B. W. Kernighan, AMPL: A Modeling Language for Mathematical Programming, Duxbury Press / Brooks/Cole Publishing Company, 1993. http://www.ampl.com/cm/cs/what/ampl/

[4] GAMS/Convert 4.0 User Notes, PDF-document, 2002. http://www.gams.com/docs/contributed/convert.pdf

[5] GAMS World, WWW-document, 2002. http://www.gamsworld.org

[6] GLOBAL Library, WWW-document, 2002. http://www.gamsworld.org/global/globallib.htm

[7] GMS2XX Translator, WWW-document, 2002. http://www.gamsworld.org/translate.htm

[8] N. I. M. Gould, D. Orban and Ph.L. Toint, CUTEr, a constrained and unconstrained testing environment, revisited, WWW-document, 2001. http://cuter.rl.ac.uk/cuter-www/problems.html

[9] R. B. Kearfott, Rigorous Global Search: Continuous Problems, Kluwer, Dordrecht 1996. www.mscs.mu.edu/~globsol

[10] B. A. Murtagh and M. A. Saunders, MINOS 5.4 User's Guide, Report SOL 83-20R, Systems Optimization Laboratory, Stanford University, December 1983 (revised February 1995). http://www.sbsi-sol-optimize.com/Minos.htm

[11] H. Mittelmann, Benchmarks. WWW-document, 2002. http://plato.la.asu.edu/topics/benchm.html

[12] J. D. Pinter, Global Optimization in Action, Kluwer, Dordrecht 1996. http://www.dal.ca/~jdpinter/l_s_d.html

[13] M. Tawarmalani and N. V. Sahinidis, Convexification and global optimization in continuous mixed-integer nonlinear programming, Kluwer, Dordrecht 2002. http://archimedes.scs.uiuc.edu/baron/baron.html

[14] B. Vanderbei, Nonlinear Optimization Models, WWW-document. http://www.orfe.princeton.edu/~rvdb/ampl/nlmodels/

[15] P. Van Hentenryck, L. Michel and Y. Deville, Numerica. A Modeling Language for Global Optimization, MIT Press, Cambridge, MA 1997.

Appendix. The Standard Unit Time

The **standard unit time** is defined as the cpu time needed to carry out the C++ program shekel5.cpp available from

> http://www.mat.univie.ac.at/~neum/glopt/coconut/

compiled without any optimization or debugging option; in particular with

> g++ -o shekel5.x ./shekel5.cpp

on UNIX/LINUX systems and with

> cl -o shekel5.exe ./shekel5.cpp

on MS/DOS systems. The program evaluates the 4-dimensional Shekel5 test function at 10^8 specified points. (Reflecting the increased speed of modern computers, this is a factor of 10^5 larger than the standard unit time used in the first global optimization benchmark by DIXON & SZEGÖ [1] in 1974.)

Quality Assurance and Global Optimization

Michael R. Bussieck[1], Arne Stolbjerg Drud[2],
Alexander Meeraus[1], and Armin Pruessner[1]

[1] GAMS Development Corp.
1217 Potomac Street, NW, Washington, DC 20007, USA
{MBussieck,AMeeraus,APruessner}@gams.com
http://www.gams.com
[2] ARKI Consulting & Development A/S
Bagsvaerdvej 246A, 2880 Bagsvaerd, Denmark
ADrud@arki.dk
http://www.conopt.com

Abstract. GAMS Development and ARKI Consulting use Quality Assurance (QA) as an integral part of their software development and software publishing process. Research and development in the global optimization area has resulted in promising implementations. Initiated by customer demand, we have been adding three different global codes, BARON, LGO, and OQNLP, to our portfolio of nonlinear optimization solvers. As part of our QA effort towards the integration and testing of these new global solvers an open architecture for reliability and performance testing of mixed-integer nonlinear optimization codes has been released. This open testing framework has been placed in the public domain (www.gamsworld.org) to serve our customers, and researchers in general, by making reproducible tests a practical proposition. We give examples illustrating the quality assurance process for obtaining performance results for local and global nonlinear and mixed-integer nonlinear programming solvers, using the existing framework tools described in this article.

1   Introduction

The research and development efforts in the area of global (nonlinear) optimization codes have increased significantly in recent years, and a number of large-scale implementations have been successfully deployed in specialized research environments. To make these systems more widely available to users of optimization system, the global solvers need to be integrated into a modeling system that manages problem formulation and guides the solution process. In this paper we report on the quality assurance aspects of transforming a research product into a reliable commercial product. When introducing new solver technology into the marketplace, there are additional difficulties besides technical problems, due to the inherent risk to the customer using commercially unproven technology. Furthermore, the potential failure of a single global optimization code today can reduce the confidence level in future global optimization solvers.

C. Bliek et al. (Eds.): COCOS 2002, LNCS 2861, pp. 223–238, 2003.

This risk is comparable to the introduction of mixed-integer linear (MIP) solvers in the 1960s and 1970s.

To address some of the problems associated with introducing a new technology into a mature modeling environment, we have decided to share our internal quality assurance tools and make them available to everyone. Quality Assurance (QA), the process of assuring the quality of one organization's outcome, is in our case means to assure our customers of reliable, state-of-the-art technology. Although QA has become an essential component in most industrial and commercial undertakings, however, it has been more or less ignored by the Mathematical Programming (MP) community. This should not come as a surprise, since the main market for MP is the academic literature. Running the risk (again) of annoying some our colleagues, we would like to draw some historical analogies.

We, the optimization modelers, are in a transition from a slow and inefficient cottage industry, similar to manufacturing just prior to the Industrial Revolution. Lone experts during this period painstakingly manufactured hand-tooled, customized items. We are now transitioning to an industrial, customer-driven environment with standards, interchangeable parts, and low-cost, highly-reliable components. This transition is typical for engineering and problem solving, activities trying to maximize the use of existing parts and components and breaking larger problems into smaller, more tractable items. Standards are essential to increase the reliability and effectiveness of our products. Customers demand quality, which is itself a fuzzy concept (a bit like beauty being in the eye of the beholder). Quality Management and Standards have developed into an industry by itself, and ISO 9000, the International Standard Organization's Quality Management System Standard, has become the mainstay of many industries.

Another way to look at this positive development is the shift of the central theme in practical modeling. The first phase was dominated by *algorithmic issues*. Problem representations were defined by algorithms, and performance testing focused on detailed algorithmic issues. A good example is the seminal paper by Crowder, Dembo and Mulvey [3] on computational experiments. The second phase is dominated by *data and model representation* issues. This phase has resulted in a number of algebra-based modeling systems pioneered by LINDO [18], GAMS [1] and AMPL [7] (in that chronological order). We are now transitioning into a third phase dominated by *real-life problem solving*. There has been a permanent shift from a scientific, supply-driven regime into a market-oriented, user demand-driven business environment. In this environment, quality has become a central issue, as in any other industry.

In this paper we discuss some of the elements of an effective QA framework. In §2 we give a brief overview of the history of the GAMS modeling language and discuss the technical principles of modeling languages. We also discuss general global optimization and some issues associated with the quality assurance of global solvers. In §3 we describe the necessary tools for effective quality assurance testing, and in §4 we stress the importance of open testing architectures. In §5 we give examples implementing the framework we have described, and finally, in §6, we draw conclusions.

2   Modeling Languages and Global Optimization Codes

The use of modeling languages has greatly simplified the solution of large-scale practical problems, both in academia and in industry. Because of its large and wide-ranging client base, GAMS has a deep impact on nonlinear programming (NLP) solver technology, and thus a responsibility for providing high quality and reliable commercial (local and global) NLP solvers.

In this section we describe some of the principles of modeling languages and some issues specific to global optimization codes which need to be addressed by any effective quality assurance framework.

2.1   Basic Technical Principles

Throughout the evolution as a company, GAMS has adhered to three basic technical principles:

Separation of Model and Solution Methods: Separating the model from the various solution methods incorporated within that model ensures that the user is not locked into any particular method, and can switch rapidly between models at no additional cost. For example, users can seamlessly switch local and global solvers. This separation of model and solution method is now accepted as a standard approach to general modeling.

Computing Platform Independence: Platform independence ensures immediate application on any user's configuration, and eliminates conversion costs. Although most modeling systems in the commercial world are on the Windows platform, customer decisions are often influenced by availability across platforms.

Multiple Solvers, Platforms, and Model Types: To create the most flexible general model, GAMS' software incorporates solvers available from both the academic and the commercial worlds. Multiple solver, platform, and model type flexibility enables the user to tackle problems involving linear programs (LP), MIP, NLP, mixed-integer nonlinear programs (MINLP), mixed complementarity problems (MCP), mathematical programs with equilibrium constraints (MPEC), stochastic programming, and models written in MPSGE, a language for solving computable general equilibrium(CGE) models.

2.2   Global Optimization Principles

Most practical models involving nonlinearities are developed in a modeling language, and most NLP and MINLP solvers are linked to a modeling system. In our modeling system, GAMS, we try to improve the reliability of NLP modeling and thereby reduce the risk for our customers by offering a *portfolio* of NLP solvers (BARON, CONOPT, LGO, MINOS, MOSEK, OQNLP, PATHNLP, SNOPT - see [8] and the references therein) implementing a variety of different algorithms (interior point, GRG, SLP, SQP). This approach improves the probability of solving our customers' models.

Solving MINLP problems involves the sequential solution of a large number of NLP sub-problems (either in a branch-and-bound, extended cutting plane, or outer approximation algorithm), which increases the chance of failure of the overall algorithm. Our MINLP solvers DICOPT and SBB have been built around this *chance of failure*. By enabling these solvers to access any NLP sub-solver in our portfolio we effectively minimize the chance of failure. The use of a multi-solver architecture helps in cases of solver failure, and overcomes the weakness inherent in local optimization codes (*local solutions*, in particular *local infeasibility*). Nonlinear modelers have coped with this weakness by providing good starting points, and have implemented their own multi-start methods (usually with random points) at a modeling language level.

The incentive structure for developers of global optimization algorithms such as BARON [19], LGO [17], and OQNLP [20] in the academic environment is very different from that faced in a commercial setting. Algorithms in the academic world are operated in *expert mode* by the developers themselves, and are often highly specialized to meet the requirements of an abstract problem class. Commercial solvers, however, are deployed by users who have no interest in the algorithm itself but want to solve their business problem by using the algorithm in a *black box mode*. Instead of delivering extraordinary performance requiring substantial use of specific options or *code tweaking*, a commercial solver has to work reliably with decent performance in all cases using *default settings*. In case of algorithmic failure, the solver has to terminate gracefully and issue suggestions on how to overcome the failure.

Since its introduction, the GAMS system has provided nonlinear modeling to a wide audience of academic and commercial users. Global solvers can be almost seamlessly integrated into the portfolio of nonlinear solvers, offering a new service to our customers and opening a new market to global solver providers. The risk for our customers of investing in new solver technology can be reduced by providing access to quality assurance results for global solvers and a plug-compatible interface that allows painless transition from local solvers to global ones.

Global Optimization Specific Issues Global optimization requires special attention to several specific issues. These issues include:

Termination Criteria: While the termination criteria for local solvers is concise (satisfying the Karush-Kuhn Tucker conditions), the stopping criteria for global codes is ambiguous. Mutli-start methods use different starting points to converge to different local optima, thereby maximizing the probability that one of the local solutions is indeed the global optima. Unfortunately, the algorithm cannot determine precisely if a current local optima is the global optima.

Problem Bounds: Global solvers generally require modification of the original problem. In particular, global solvers can usually only guarantee global optimality if the problem is bounded for both variables and expressions (bounding box principle).

Limited Algebra: Some global optimization solvers do not support all functions. In particular, many global solvers cannot handle black-box-type functions, where only function evaluations are returned. In particular, many require detailed knowledge of the algebra involved in the function itself.

Solution Quality Metrics: While for linear and local nonlinear solvers robustness and efficiency metrics are sufficient in characterizing solver performance and assure quality, for global optimization solvers quality of solution is another descriptive metric of performance. When analyzing global solver performance, higher solution quality is often obtained at a cost of efficiency. Thus, users should consider if feasibility of the (local) solution and efficiency or global optimality at a cost of efficiency is the primary criteria.

2.3   Quality Assurance

Although reproducibility of test results has been accepted as a critical step in most scientific fields, computational results in our field can rarely be reproduced. Limited access to test cases, non-reproducible methods for collecting results, and non-standard analysis tools seem to prevent the low-cost replication of such results by an independent auditor. Based on our experience, we rank the *replication of quality assurance results* as the most critical factor for establishing a new solver technology in the commercial world.

In the next section we discuss the necessary components for an effective quality assurance framework.

3   Effective Quality Assurance Testing

The key ingredients for effective testing are diverse *test cases*, efficient *data collection* tools, and automated *data analysis* tools.

The choice of test problems for benchmarks is difficult and inherently subjective. While there is no consensus on choosing appropriate models, the use of standard model libraries is important to ensure that testing is performed with a wide selection of models from diverse application areas. Consider that a particular solver may be fine-tuned for specific applications. The use of diverse standard model libraries *reduces the risk of bias* if a solver is fine-tuned for a specific family of models and allows more objective cross-comparisons of various solvers.

Most optimization engines or solvers output model statistics, objective function, resource time and general solve information. Many benchmarks and performance analyses involve either manually extracting the necessary information from the log output or writing solver and optimization engine-specific scripts to parse the output and extract the data to be analyzed. This can be cumbersome and is error prone and generally must be tailored to the specific engine or solver. In order to simplify the quality assurance process, data collection must be *automated*.

Finally, few standard performance metrics exist and the reproducibility of performance tests is often not a practical proposition. By introducing automated tools and standard performance measurements, we enable the quality assurance process to be *inexpensive, efficient, and reproducible*.

3.1 Test Cases

Testing global and local optimization solvers requires access to a collection of test models, including toy models, academic application models and, most importantly, for our purposes, commercial application models (which are in general difficult to collect due to the proprietary nature of most commercial models). While academic application models are useful in their own right, they do not always address the same problem types and model structure a commercial model may. In order to assure a new solver technology's quality in the commercial world, any practical model library should contain commercial application models as well.

Our publicly available collection of NLP and MINLP models is:

GAMS Model Library: `http://www.gams.com/modlib/modlib.htm`, with more than 250 models from over 18 application areas.
GLOBALLib: `http://www.gamsworld.org/global/globallib.htm`, with more than 390 scalar NLP models.
MINLPLib: `http://www.gamsworld.org/minlp/minlplib.htm`, with about 180 scalar MINLP models. Also see [2].
MPECLib: `http://www.gamsworld.org/mpec/mpeclib.htm`, which produces over 10,000 NLP models .

New models are added continuously to these model libraries. In order to add commercial models with proprietary data, we make use of the CONVERT utility.

Translating Models Using CONVERT For commercial application models, where proprietary data should be hidden, GAMS provides the CONVERT tool, which translates a model into *scalar* format. This hides all proprietary parts of a model and makes public access to customer models possible.

Modeling languages such as GAMS or AMPL have a rich syntax that is usually based on sets and indexed variables, equations, and parameters. This syntax, and the corresponding structure in the model and the data, is very useful for the model developer. However, usually such structures are not used by the solvers. Most solvers see the world as a list of variables, X_1 to X_n, a list of equations or constraints, E_1 to E_m, and the relationship between these variables and equations, as represented in some form. As long as the model seen by the solver remains unchanged, it is therefore acceptable for a translator to remove the structure. The GAMS translator CONVERT transforms models into a very simple, internal scalar format. This internal format can then be written out in many different formats. With GAMS as output format, the scalar model consists of the following:

- Declarations of the variables, with extra declarations for the subsets of positive, integer, or binary variables,
- Declarations of the equations,
- The symbolic form of these equations, and
- Assignment statements for non-default bounds and initial values.

All operations involving sets are unrolled, and all expressions involving parameters are evaluated and replaced by their numerical values. Since there are no sets or indexed parameters in the scalar models, most of the differences between modeling systems have disappeared. Therefore, the GAMS format can be easily transformed into another language's format. For example, in AMPL the keyword "var" is used instead of "Variable," bounds and initial values are written using a different format, the equation declarations are missing, the equation definitions start with "subject to E_i" instead of "E_i..", and a few operators are named differently.

There are a few cases where a model cannot be translated into a particular language because special functions (e.g. `errorf`) or variable types (e.g. `SemiCont`) are not available in that language. GAMS models have an objective variable rather than an objective function. If there is one defining equation for a given variable, CONVERT will eliminate the objective variable and will introduce a real objective function for formats that support objective functions (e.g. AMPL). For more details on CONVERT see [2].

Other issues which impact the translation include scaling of the model and presolve capabilities by the particular solver or optimization engine. In general, long term strategies should include a mathematical programming *standard scalar format*, which allows automorphic translation to and from this format. This has been accomplished for linear model with the long-accepted standard MPS format [9] and recently work focused on a broader set of models has been done by introducing a format based on XML [13].

3.2   Data Collection Tools

Reproducible data production and collection requires an automated system that provides information about the testing environment (version of software, hardware, etc.), and status, performance, and exception information for the individual test case. Such a system is automatically available in GAMS through the *trace facility* for all solvers connected. The trace facility provides access to model statistics, non-default input options, and solver and solution statistics and writes information conveniently to an ASCII interface. Table 1 shows the possible trace file column headers and the associated data.

Table 1. GAMS Trace Facility ASCII Interface (current trace file data)

Heading	Description
Filename	GAMS model filename
Modeltype	LP, MIP, NLP, etc.
Solvername	
NLP def.	Default NLP solver
MIP def.	Default MIP solver
Juliantoday	start day/time of job
Direction	0=min, 1=max
Equnum	Total number of equations
Varnum	Total number of variables
Dvarnum	Total number of discrete variables
Nz	number of nonzeros
Nlnz	number of nonlinear nonzeros
Optfile	1=optfile included, 0=none
Modelstatus	GAMS model status
Solverstatus	GAMS solver status
Obj	Value of objective function
Objest	Estimate of objective function
Res used	Solver resource time used (sec)
Iter used	Number of solver iterations
Dom used	Number of domain violations
Nodes used	Number of nodes used

The resulting data can easily be analyzed for example through user automation scripts, Excel spreadsheets or some of the online tools available as part of Performance World: http://www.gamsworld.org/performance.

3.3 Data Analysis Tools

In addition to custom programs for processing testing results, we have implemented a variety of performance measurement tools. Together with experts in the field of benchmarking at Performance World, we have also extended and developed new ways to present testing results in an easy and consistent way.

In particular, the tools as part of the *PAVER server* (Performance Analysis and Visualization for Efficient Reproducibility) [15], accessible online at

http://www.gamsworld.org/performance/paver

can be used for performance analysis of performance data collected in trace files in an automated fashion. The server provides automation and visualization tools for automatically generating HTML-based reports of submitted trace files and gives information on robustness, efficiency and solver quality of solution. Users can submit up to 8 trace files (each containing performance data for a particular

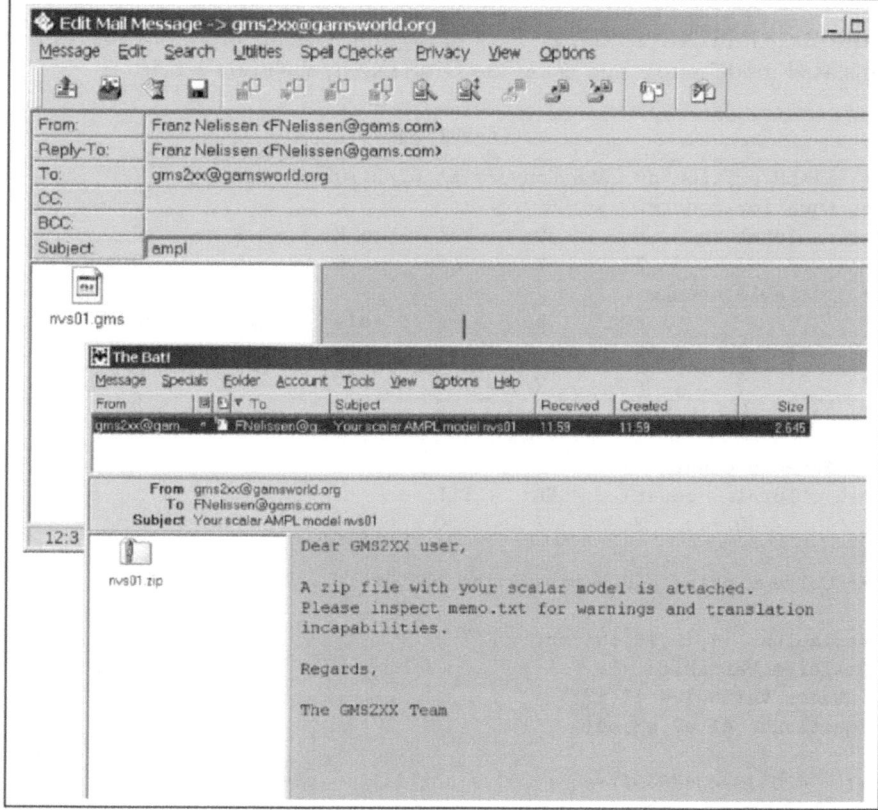

Fig. 1. GMS2XX/CONVERT E-mail submission and results

solver) online, and the server will automatically generate quality and performance reports and do cross-comparisons of all solvers.

Test Cases: The CONVERT tool is also available as an e-mail based translation service (GMS2XX) at GAMS World

http://www.gamsworld.org/translate.htm

to facilitate the translation of GAMS models into other modeling languages, such as AMPL, BARON, CplexLP, CplexMPS, LGO, LINGO, and MINOPT. Users submit their model as an e-mail attachment to gms2xx@gamsworld.org with the translation language listed in the subject line. The translated model is then returned via e-mail attachment. Figure 1 shows the process of converting the MINLPLib model nvs01.gms from GAMS to AMPL via the GMS2XX translation service. Table 2 shows the original model nvs01.gms and the converted AMPL model ampl.mod.

Data Collection Tools: The trace facility has an open API and allows ASCII trace file importing and exporting. Non-GAMS users can therefore produce

Table 2. GMS2XX Example: Part of original GAMS Model `nvs01.gms` and translated model `ampl.mod` using the GMS2XX translation service

nvs01.gms

```
*   MINLP written by GAMS Convert at 02/21/03 13:01:13
*   Equation counts
*      Total        E        G        L        N        X        C
*        4           2        2        0        0        0        0
*   Variable counts
*                    x        b        i      s1s      s2s       sc       si
*      Total      cont   binary integer     sos1     sos2    scont     sint
*        4           2        0        2        0        0        0        0
*   FX   0           0        0        0        0        0        0        0

*   Nonzero counts
*      Total     const       NL      DLL
*        10         3         7        0
*
*   Solve m using MINLP minimizing objvar;

Variables  i1,i2,x3,objvar;
Positive Variables x3;
Integer Variables i1,i2;
Equations  e1,e2,e3,e4;

e1.. 420.169404664517*sqrt(900 + sqr(i1)) - x3*i1*i2 =E= 0;
e2.. - x3 =G= -100;
e3.. 296087.631843*(0.01+0.0625*sqr(i2))/(7200 + sqr(i1))-x3 =G= 0;
```

ampl.mod

```
#  MINLP written by GAMS Convert at 02/21/03 14:03:38
#
#  Reformulation has removed 1 variable and 1 equation

var i1 integer := 100, >= 0, <= 200;
var i2 integer := 100, >= 0, <= 200;
var x3 := 100, >= 0, <= 100;

minimize obj: 0.04712385*i2*(900 + i1^2)^0.5;

subject to

e1: 420.169404664517*sqrt(900 + i1^2) - x3*i1*i2 = 0;
e2: - x3 >= -100;
e3: (2960.87631843 + 18505.4769901875*i2^2)/(7200 + i1^2) - x3 >= 0;
```

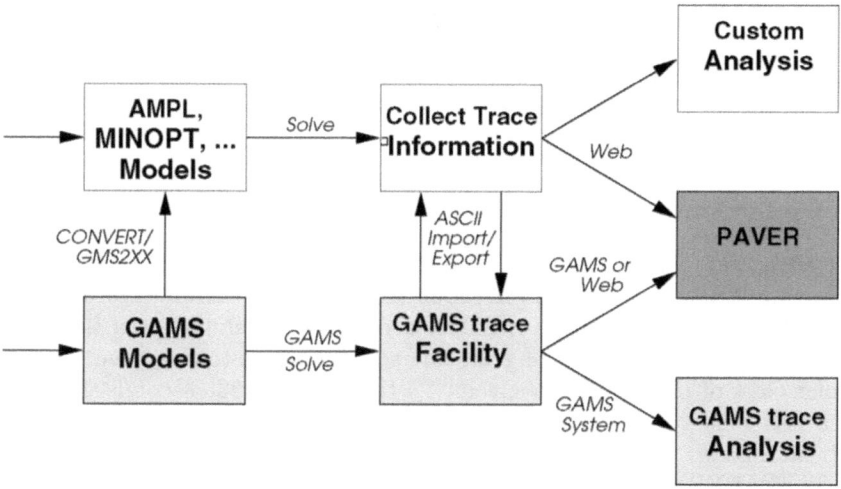

Fig. 2. Quality Assurance Process. The process is system and software independent at each phase. Models can be translated to other languages, benchmark runs performed with any optimization engine, and subsequent analysis done using web-based, GAMS, or customized tools

a trace file by any preferred method, for example by writing scripts to parse the log output from an optimization engine or manually writing a trace record.

Data Analysis Tools: All performance tools are implemented in open source GAMS programs. Free access to these programs for users without a GAMS system is guaranteed through the Web interface PAVER described previously. The server is not restricted to trace files obtained through GAMS for generating performance analysis reports but can accept data from any optimization engine providing trace-like data files.

The open architecture and steps in the quality assurance process are illustrated in Figure 2. A user with a GAMS system can solve the model and automatically capture solve and model statistics via the trace file utility. The data can be analyzed then via GAMS or the PAVER web interface. Users of other modeling systems or optimization engines can use the GMS2XX tool to convert the model to another language. Once solved with this other engine, trace files can be created either manually or through some other automated fashion. Analysis can then proceed either customized or through the PAVER web interface. Note also that the ASCII trace interface allows simple import and export of data between GAMS and other formats.

4   Examples

In this section we illustrate how to create reproducible quality assurance results and performance results for local and global NLP and MINLP solvers using the framework described previously.

4.1   NLP Example: Linear Multiplicative Models

The first example involves 50 different instances of the NLP model lmp1.gms (linear multiplicative models [12]) found in the GAMS Model Library. We used the global optimization code BARON and ran it in default mode and with five different option files, specifying varying solver options to fine tune for this particular class of problems. We specified a time limit of 300 seconds. Performance data was collected via trace files from within GAMS and we used the PAVER web submission utility to compare performance.

The PAVER results returned via e-mail give a variety of information including robustness and efficiency information, as well as quality of solution. We show the *performance profiles* results [4] which provide a measure of competitiveness of a solver.

Figure 3 shows competitiveness of each BARON run with various options if we are interested only in solver efficiency. At a Time Factor of 1 (x-axis), the graph shows the percentage of models solved the fastest by a particular solver. In particular, BARON with option 6 (BARON op6) solves 65% of the models the fastest. As Time Factor goes to ∞, the graph gives solver robustness information, indicating the percentage of models a solver can solve at all given any amount of time. In this case BARON default has the highest probability of success at roughly 95%. Also note the graph associated with CAN_SOLVE, which shows the probability of success that any one of the six "solvers" can solve the models.

4.2   MINLP Example: Models from Gupta and Ravindran

In this example we choose 24 MINLP models from [10], available as part of the MINLPLib. We ran all available GAMS MINLP solvers (BARON, DICOPT, OQNLP, SBB). We also ran MINLP, a branch and bound code based on FILTER by Fletcher and Leyffer [14]. The solver MINLP was run remotely through the Network-Enabled Optimization Server (NEOS) [5], [11], [16], an online server for solving optimization problems. All solvers were run using a time limit of 20 seconds. All performance information for GAMS solvers was collected in trace files. For the MINLP solver run remotely through the NEOS server, we wrote a script to parse the output and create a trace file.

In order to evaluate solver performance we wrote a GAMS script to compare objective value information. The script also computes absolute and relative gaps and compares CPU time. The script first reads the trace file with the performance data, computes statistical information and then writes the output to a standard text file. The program is listed below.

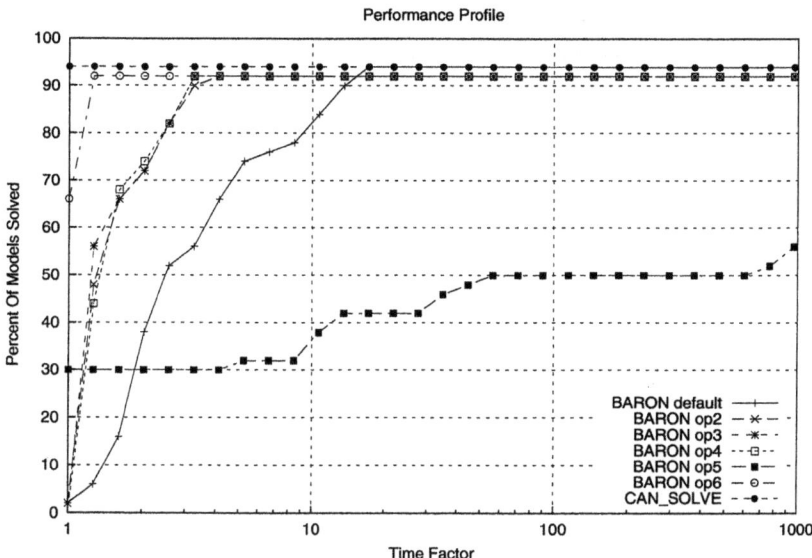

Fig. 3. PAVER Performance Profiles for LMP1 models. Graphs show measure of competitiveness of each "solver" in terms of efficiency. BARON with option 6 is the most efficient (see the profile for `Time Factor = 1`, solving 65% of the models the fastest. In terms of probability of success, BARON default has the highest rating. Roughly 95% of the models can be solved by BARON default given any amount of time

MINLP Example: GAMS Analysis Script

```
$set  col     julian,dir,eqnum,varnum,dvarnum,nz,nlnz,opt,modelstat
$set  col     %col%,solvestat,obj,object,res,iter,dom,nodes
Set   col     /%col%/,
      c(col) /modelstat,solvestat,obj,object,res,nodes/

Alias(u1,u2,*);
$eolcom ,#

*=== Read in trace file
Table tracedata(*,*,col)
$ondelim
modelname,solvername,%col%
$offlisting
$call cat trace.trc minlpbb.csv | cut -d, -f1,3,6- > trace.tmp
$include trace.tmp
$onlisting
$offdelim
;
```

```
*=== Extract driving sets
Set modelname, solvername;
loop((u1,u2,)$tracedata(u1,u2,'julian'),
  modelname(u1)  = yes;
  solvername(u2) = yes;
);

parameter srep(u1,u2,col)
          mstat(u1,u2);

*=== Load reference solution
$gdxin minlpstat
$load mstat

*=== Select columns
srep(modelname,solvername,c) =
      tracedata(modelname,solvername,c);
srep(modelname,'Reference','obj') =
      mstat(modelname,'BestInt');

Parameter gap;
gap(modelname,solvername,'agap') =
      round(srep(modelname,solvername,'obj')
    - srep(modelname,'Reference','obj'),3);
gap(modelname,solvername,'agap')$(
      srep(modelname,solvername,'modelstat')<>1
   and srep(modelname,solvername,'modelstat')<>2
   and srep(modelname,solvername,'modelstat')<>8
                              ) = inf;
gap(modelname,solvername,'rgap') =
      gap(modelname,solvername,'agap')
    / abs(srep(modelname,'Reference','obj'));
gap(modelname,solvername,'obj')  =
      srep(modelname,solvername,'obj');
gap(modelname,solvername,'cpu')  =
      srep(modelname,solvername,'res');

display srep, gap;
```

The output created by the script shows the computed parameters srep and gap. It shows the objective function value, the absolute and relative gaps, as well as the CPU time used for a subset of models for each of the solvers in the comparison.

MINLP Example: GAMS Analysis Script

```
----     167  PARAMETER gap

                  obj      agap      rgap       cpu
NVS01.OQNLP    15.806     3.336     0.268     2.218
```

NVS01.BARON	12.470		0.110
NVS01.SBB	12.470		0.410
NVS01.DICOPT	12.470		0.035
NVS01.MINLPBB	12.470		20.000
NVS02.OQNLP	6.575	0.611	0.102	20.046
NVS02.BARON	5.964		0.080
NVS02.SBB	5.964		0.437
NVS02.DICOPT	5.964		0.346
NVS02.MINLPBB	5.964		20.000
NVS03.OQNLP	16.000		20.015
NVS03.BARON	16.000		0.060
NVS03.SBB	16.000		0.305
NVS03.DICOPT	16.000		0.069
NVS03.MINLPBB	16.000		20.000
NVS04.OQNLP	0.720		0.265
NVS04.BARON	0.720		0.050
NVS04.SBB	0.720		0.195
NVS04.DICOPT	2.120	1.400	1.944	0.227
NVS04.MINLPBB	0.720		20.000

5   Conclusions

We have addressed the necessary steps for moving global optimization codes from an academic research lab into the commercial production environment. Reproducible quality assurance testing is the key to success. We have proposed a testing framework, which includes model collections, data collection tools and data analysis tools, and allow seamless testing inside the GAMS system, yet remains open for outside use by non-GAMS customers and researchers in general. We offer these models and tools along with our open invitation to use them as a means to strengthen collaboration among all groups interested in quality assurance for global optimization.

References

[1] Brooke, A., Kendrick, D., Meeraus, A.: GAMS: A User's Guide, The Scientific Press, Redwood City, California (1988)
[2] Bussieck, M. R., Drud, A. S., Meeraus, A.: MINLPLib - A Collection of Test Models for Mixed-Integer Nonlinear Programming, Informs J. Comput., **15** (1) (2003) 114–119
[3] Crowder, H., Dembo, R. S., Mulevy, J. M.: On Reporting Computational Experiments with Mathematical Software, ACM Transactions of Mathematical Software, **5** (2) (1979) 193–203
[4] Dolan, E. D., Moré, J. J.: Benchmarking Optimization Software with Performance Profiles, Math. Programming, **91** (2002) 201–213
[5] Dolan, E. D.: The NEOS Server 4.0 Administrative Guide, Technical Memorandum ANL/MCS-TM-250, Mathematics and Computer Science Division, Argonne National Laboratory (2001)

[6] Dolan, E. D., Moré, J. J.: Benchmarking Optimization Software with COPS, Technical Memorandum ANL/MCS-TM-246, Argonne National Laboratory, Argonne, Illinois (2000)

[7] Fourer, R., Gay, D. M.: AMPL: A Modeling Language for Mathematical Programming, The Scientific Press, Redwood City, California (1993)

[8] GAMS Development Corp.: GAMS - The Solver Manuals, GAMS Development Corp., Washington, D. C. (2003)

[9] Gay, D. M.: Electronic Mail Distribution of Linear Programming Test Problems, Mathematical Programming Society COAL Newsletter (1985)

[10] Gupta, O. K. Ravindran, A.: Branch and Bound Experiments in Convex Nonlinear Integer Programming, Management Science, **13** (1985) 1544–1546

[11] Gropp, W., Moré, J. J.: Optimization Environments and the NEOS Server, In: M. D. Buhmann and A. Iserles (Eds.), Approximation Theory and Optimization, Cambridge University Press, Cambridge (1997) 167–182

[12] Konno, H., Kuno, T.: Linear Multiplicative Programming, Math. Prog., **56** (1992) 51–64

[13] Kristjansson, B.: Optimization Modeling in Distributed Applications: How New Technologies such as XML and SOAP allow OR to provide web-based Services, INFORMS Roundtable, Savannah, Georgia (2001)

[14] Leyffer, S.: User Manual for MINLP_BB, University of Dundee Numerical Analysis Report NA/XXX (1999)

[15] Mittelmann, H., Pruessner, A.: A Server for Automated Performance Analysis and Benchmarking of Optimization Software, submitted (2003)

[16] NEOS: http://www-neos.mcs.anl.gov/ (1997)

[17] Pintér, J. D: LGO - A Model Development System for Continuous Global Optimization. User's Guide. (Current revised edition.), Pintér Consulting Services, Halifax, Nova Scotia (2002)

[18] Schrage, L. S.: LINDO - An Optimization Modeling System, Scientific Press series; Fourth ed. Boyd and Fraser, Danvers, Massachussets (1991)

[19] Tawarmalani, M., Sahinidis, N. V.: Convexification and Global Optimization in Continuous and Mixed-Integer Nonlinear Programming: Theory, Algorithms, Software, and Applications, Kluwer Academic Publishers, Dordrecht (2002)

[20] Ugray, Z., Lasdon, L., Plummer, J., Glover, F., Kelly, J., Marti, R.: A Multistart Scatter Search Heuristic for Smooth NLP and MINLP Problems, INFORMS J. Comp., to appear (2002)

Author Index

Lecture Notes in Computer Science

For information about Vols. 1–2798
please contact your bookseller or Springer-Verlag

Vol. 2838: N. Lavrač, D. Gamberger, L. Todorovski, H. Blockeel (Eds.), Knowledge Discovery in Databases: PKDD 2003. Proceedings, 2003. XVI, 508 pages. 2003. (Subseries LNAI).

Vol. 2839: A. Marshall, N. Agoulmine (Eds.), Management of Multimedia Networks and Services. Proceedings, 2003. XIV, 532 pages. 2003.

Vol. 2840: J. Dongarra, D. Laforenza, S. Orlando (Eds.), Recent Advances in Parallel Virtual Machine and Message Passing Interface. Proceedings, 2003. XVIII, 693 pages. 2003.

Vol. 2841: C. Blundo, C. Laneve (Eds.), Theoretical Computer Science. Proceedings, 2003. XI, 397 pages. 2003.

Vol. 2842: R. Gavaldà, K.P. Jantke, E. Takimoto (Eds.), Algorithmic Learning Theory. Proceedings, 2003. XI, 313 pages. 2003. (Subseries LNAI).

Vol. 2843: G. Grieser, Y. Tanaka, A. Yamamoto (Eds.), Discovery Science. Proceedings, 2003. XII, 504 pages. 2003. (Subseries LNAI).

Vol. 2844: J.A. Jorge, N.J. Nunes, J.F. e Cunha (Eds.), Interactive Systems. Proceedings, 2003. XIII, 429 pages. 2003.

Vol. 2846: J. Zhou, M. Yung, Y. Han (Eds.), Applied Cryptography and Network Security. Proceedings, 2003. XI, 436 pages. 2003.

Vol. 2847: R. de Lemos, T.S. Weber, J.B. Camargo Jr. (Eds.), Dependable Computing. Proceedings, 2003. XIV, 371 pages. 2003.

Vol. 2848: F.E. Fich (Ed.), Distributed Computing. Proceedings, 2003. X, 367 pages. 2003.

Vol. 2849: N. García, J.M. Martínez, L. Salgado (Eds.), Visual Content Processing and Representation. Proceedings, 2003. XII, 352 pages. 2003.

Vol. 2850: M.Y. Vardi, A. Voronkov (Eds.), Logic for Programming, Artificial Intelligence, and Reasoning. Proceedings, 2003. XIII, 437 pages. 2003. (Subseries LNAI).

Vol. 2851: C. Boyd, W. Mao (Eds.), Information Security. Proceedings, 2003. XI, 443 pages. 2003.

Vol. 2852: F.S. de Boer, M.M. Bonsangue, S. Graf, W.-P. de Roever (Eds.), Formal Methods for Components and Objects. Proceedings, 2003. VIII, 509 pages. 2003.

Vol. 2853: M. Jeckle, L.-J. Zhang (Eds.), Web Services – ICWS-Europe 2003. Proceedings, 2003. VIII, 227 pages. 2003.

Vol. 2854: J. Hoffmann, Utilizing Problem Structure in Planning. XIII, 251 pages. 2003. (Subseries LNAI)

Vol. 2855: R. Alur, I. Lee (Eds.), Embedded Software. Proceedings, 2003. X, 373 pages. 2003.

Vol. 2856: M. Smirnov, E. Biersack, C. Blondia, O. Bonaventure, O. Casals, G. Karlsson, George Pavlou, B. Quoitin, J. Roberts, I. Stavrakakis, B. Stiller, P. Trimintzios, P. Van Mieghem (Eds.), Quality of Future Internet Services. IX, 293 pages. 2003.

Vol. 2857: M.A. Nascimento, E.S. de Moura, A.L. Oliveira (Eds.), String Processing and Information Retrieval. Proceedings, 2003. XI, 379 pages. 2003.

Vol. 2858: A. Veidenbaum, K. Joe, H. Amano, H. Aiso (Eds.), High Performance Computing. Proceedings, 2003. XV, 566 pages. 2003.

Vol. 2859: B. Apolloni, M. Marinaro, R. Tagliaferri (Eds.), Neural Nets. Proceedings, 2003. X, 376 pages. 2003.

Vol. 2860: D. Geist, E. Tronci (Eds.), Correct Hardware Design and Verification Methods. Proceedings, 2003. XII, 426 pages. 2003.

Vol. 2861: C. Bliek, C. Jermann, A. Neumaier (Eds.), Global Optimization and Constraint Satisfaction. Proceedings, 2002. XII, 239 pages. 2003.

Vol. 2862: D. Feitelson, L. Rudolph, U. Schwiegelshohn (Eds.), Job Scheduling Strategies for Parallel Processing. Proceedings, 2003. VII, 269 pages. 2003.

Vol. 2863: P. Stevens, J. Whittle, G. Booch (Eds.), «UML» 2003 – The Unified Modeling Language. Proceedings, 2003. XIV, 415 pages. 2003.

Vol. 2864: A.K. Dey, A. Schmidt, J.F. McCarthy (Eds.), UbiComp 2003: Ubiquitous Computing. Proceedings, 2003. XVII, 368 pages. 2003.

Vol. 2865: S. Pierre, M. Barbeau, E. Kranakis (Eds.), Ad-Hoc, Mobile, and Wireless Networks. Proceedings, 2003. X, 293 pages. 2003.

Vol. 2867: M. Brunner, A. Keller (Eds.), Self-Managing Distributed Systems. Proceedings, 2003. XIII, 274 pages. 2003.

Vol. 2868: P. Perner, R. Brause, H.-G. Holzhütter (Eds.), Medical Data Analysis. Proceedings, 2003. VIII, 127 pages. 2003.

Vol. 2869: A. Yazici, C. Şener (Eds.), Computer and Information Sciences – ISCIS 2003. Proceedings, 2003. XIX, 1110 pages. 2003.

Vol. 2870: D. Fensel, K. Sycara, J. Mylopoulos (Eds.), The Semantic Web - ISWC 2003. Proceedings, 2003. XV, 931 pages. 2003.

Vol. 2871: N. Zhong, Z.W. Raś, S. Tsumoto, E. Suzuki (Eds.), Foundations of Intelligent Systems. Proceedings, 2003. XV, 697 pages. 2003. (Subseries LNAI)

Vol. 2875: E. Aarts, R. Collier, E. van Loenen, B. de Ruyter (Eds.), Ambient Intelligence. Proceedings, 2003. XI, 432 pages. 2003.

Vol. 2876: M. Schroeder, G. Wagner (Eds.), Rules and Rule Markup Languages for the Semantic Web. Proceedings, 2003. VII, 173 pages. 2003.

Vol. 2877: T. Böhme, G. Heyer, H. Unger (Eds.), Innovative Internet Community Systems. Proceedings, 2003. VIII, 263 pages. 2003.

Vol. 2878: R.E. Ellis, T.M. Peters (Eds.), Medical Image Computing and Computer-Assisted Intervention - MICCAI 2003. Part I. Proceedings, 2003. XXXIII, 819 pages. 2003.

Vol. 2879: R.E. Ellis, T.M. Peters (Eds.), Medical Image Computing and Computer-Assisted Intervention - MICCAI 2003. Part II. Proceedings, 2003. XXXIV, 1003 pages. 2003.

Vol. 2880: H.L. Bodlaender (Ed.), Graph-Theoretic Concepts in Computer Science. Proceedings, 2003. XI, 386 pages. 2003.

Vol. 2881: E. Horlait, T. Magedanz, R.H. Glitho (Eds.), Mobile Agents for Telecommunication Applications. Proceedings, 2003. IX, 297 pages. 2003.

Vol. 2885: J.S. Dong, J. Woodcock (Eds.), Formal Methods and Software Engineering. Proceedings, 2003. XI, 683 pages. 2003.

Vol. 2887: T. Johansson (Ed.), Fast Software Encryption. Proceedings, 2003. IX, 397 pages. 2003.

GPSR Compliance

The European Union's (EU) General Product Safety Regulation (GPSR)
is a set of rules that requires consumer products to be safe and our
obligations to ensure this.

If you have any concerns about our products, you can contact us on
ProductSafety@springernature.com

In case Publisher is established outside the EU, the EU authorized
representative is:

Springer Nature Customer Service Center GmbH
Europaplatz 3
69115 Heidelberg, Germany

Batch number: 09474024

Printed by Printforce, the Netherlands